W0267904

„Fortschritte der Hochpolymeren-Forschung/Advances in Polymer Science“

erscheinen zwanglos in einzeln berechneten Heften, die zu Bänden vereinigt werden.

Sie enthalten Fortschrittsberichte monographischen Charakters aus dem Gebiet der Physik und Chemie der Hochpolymeren mit ausführlichen Literaturzusammenstellungen. Sie sollen der Unterrichtung der auf diesen Gebieten Tätigen über solche Themen dienen, die in letzter Zeit besondere Aktualität gewonnen haben, bzw. die in neuerer Zeit eine lebhafte und nach literarischer Zusammenfassung verlangende Entwicklung erfahren haben.

Anschriften der Herausgeber:

Prof. Dr. J. D. Ferry, Department of Chemistry, The University of Wisconsin, Madison 6, Wisconsin/USA.

Prof. Dr. C. G. Overberger, Polytechnic Institute of Brooklyn, 333 Jay Street, Brooklyn 1, New York/USA.

Prof. Dr. G. V. Schulz, Institut für physikalische Chemie der Universität, Mainz.

Prof. Dr. A. J. Staverman, Fruinlaan 6, Leiden/Holland.

Prof. Dr. H. A. Stuart, Institut für physikalische Chemie der Universität, Mainz.

Springer-Verlag

Heidelberg	**Berlin-Wilmersdorf**
Neuenheimer Landstraße 28—30	Heidelberger Platz 3
Fernsprecher 27901	Fernsprecher 830301
Fernschreiber 04-61723	Fernschreiber 01-83319

---

2. Band **Inhaltsverzeichnis** 4. (Schluß-) Heft

Seite

Fortschr. Hochpolym.-Forsch., Bd. 2, S. 465—495 (1961)

Fluorine-Containing Polymers. II. Polytetrafluoroethylene*

By

C. A. SPERATI and H. W. STARKWEATHER, JR.

Polychemicals Department E. I. du Pont de Nemours and Co.
Du Pont Experimental Station, Wilmington, Delaware

With 11 Figures

Table of Contents

I. Introduction

Polytetrafluoroethylene represents in many respects the extreme of known polymers. Among its most notable properties are its high crystalline melting point (327° C), its high melt viscosity (about 10^{11} poises at

* For paper I see this Journal **1**, 75—113 (1958).

380° C), and its high thermal stability. At the other end of the temperature spectrum, it is known to possess ununsual toughness at temperatures

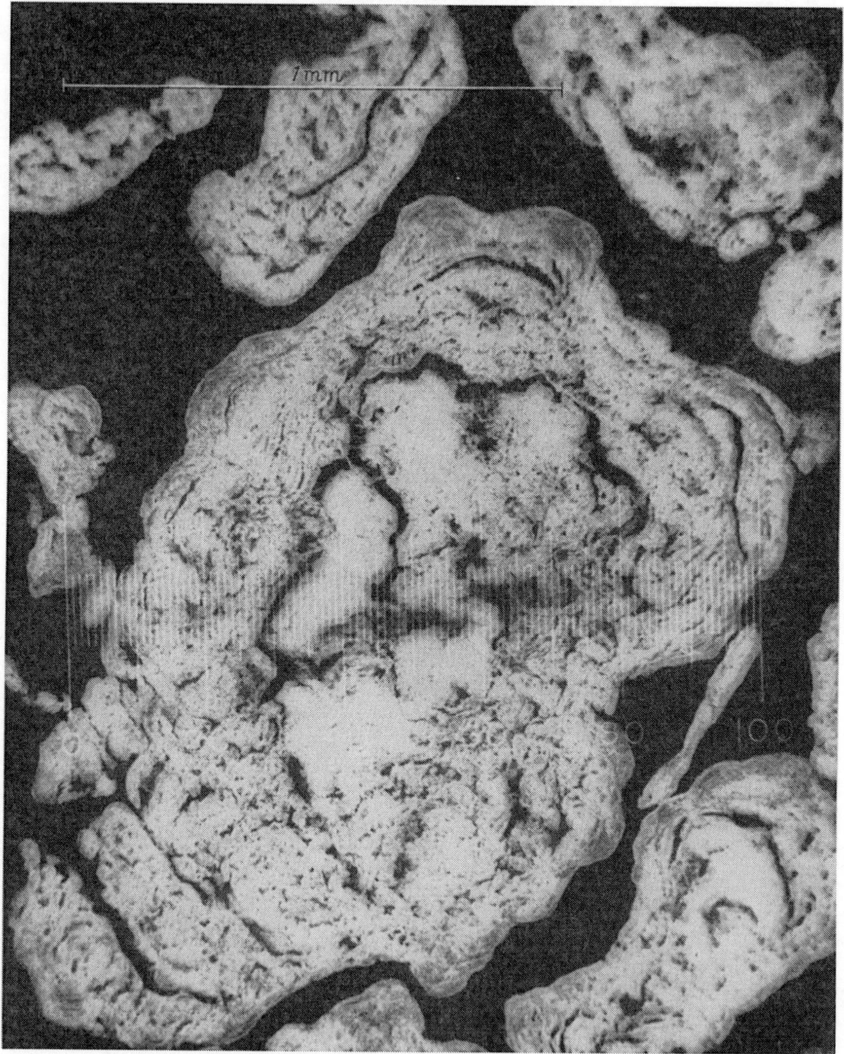

Fig. 1a and b. Micrographs of granular and dispersion particles of polytetrafluoroethylene. a) Cross-section of granular particles

as low as 4° K. The polymer is insoluble in all common solvents and highly resistant to chemical attack. It also has extremely low dielectric loss, high dielectric strength, and unique nonadhesion and antifrictional properties. While these unusual qualities offer many technological advantages, they mean that special techniques are required to determine

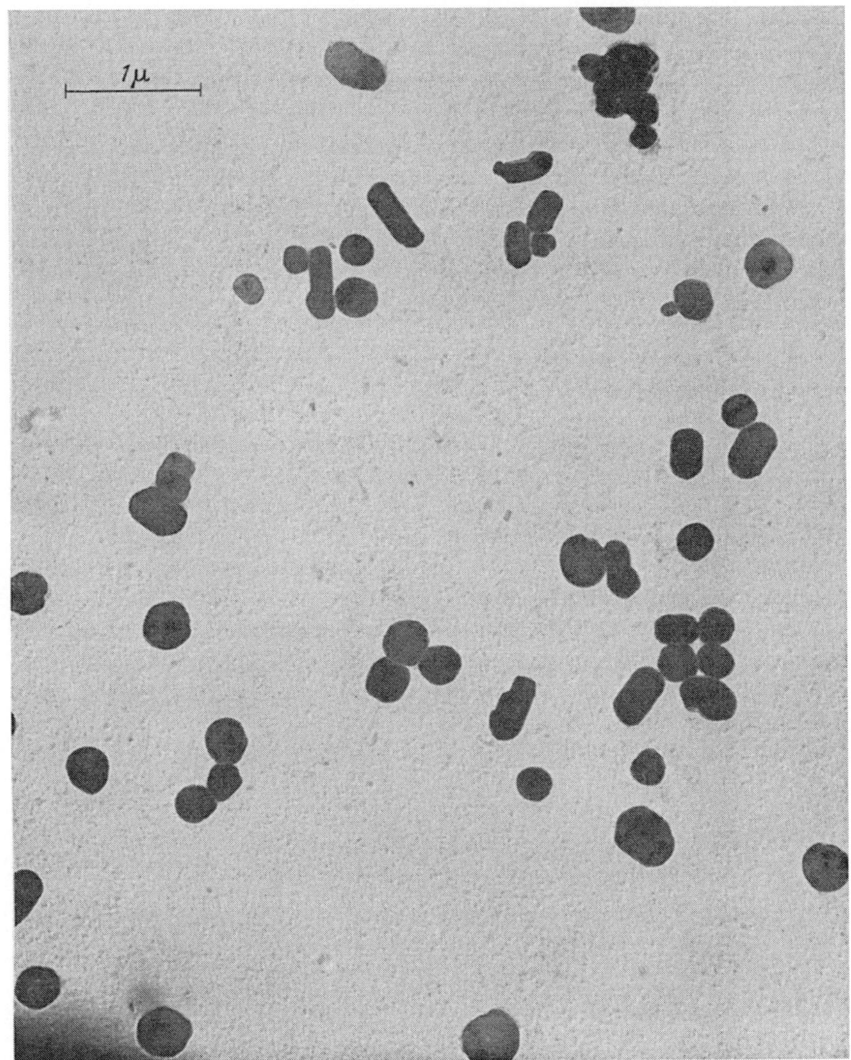

b) Dispersion particles

its molecular structure and to process it into useful shapes. This article will summarize these unusual properties and discuss the special methods which have been used to elucidate the structure of this remarkable polymer.

II. Synthesis and Polymerization of Tetrafluoroethylene

Tetrafluoroethylene is a colorless, odorless gas which boils at —76.3° C at atmospheric pressure and freezes at —142.5° C (RENFREW and LEWIS). Its critical temperature and pressure are 33.3° C and 572 psi. The heat of formation is —151.9 kcal/mole, (DUUS; BRYANT; NEUGEBAUER and MARGRAVE), and the heat of polymerization is—41.12 kcal/mole (BRYANT). The first reliable report of the preparation of this compound was in 1933 by RUFF and BRETSCHNEIDER, who decomposed tetrafluoromethane in an electric arc. Other syntheses are based on the dechlorination of sym-dichlorotetrafluoroethane (LOCKE, BRODE, and HENNE), the pyrolysis of difluorochloromethane (DOWNING, BENNING, and MCHARNESS; TORKINGTON and THOMPSON) or the decarboxylation of sodium perfluoroproprionate (HALS, REID, and SMITH).

The first report of the polymerization of tetrafluoroethylene was by PLUNKETT in 1941, who had a cylinder of tetrafluoroethylene cut open to see why the expected amount of gas was not released when the valve was opened. His perspicacity led to the discovery of an inert, white, opaque solid with a waxy feel. Various methods of polymerization were tried after the adventitious discovery and the preferred methods for polymerization now involve aqueous media and super-atmospheric pressures. Suitable initiators (HANFORD and JOYCE) include ammonium, sodium, or potassium persulfate, hydrogen peroxide, oxygen, and some organic peroxy compounds. Oxidation-reduction initiation systems involving the use of persulfate with either ferrous ion or bisulfite or the use of bisulfite with ferric ion are also useful and have been discussed by BERRY and PETERSON.

There are various kinds of polytetrafluoroethylene. One is granular polymer consisting of spongy, white particles having a median size of about 600μ. The specific surface of this polymer is on the order of 2 m^2/g (determined by nitrogen adsorption and calculations by the method of BRUNAUER, EMMETT, and TELLER). Since this specific surface area is about 1700 times the observed outer surface of the particles, these measurements confirm the porous, spongelike structure that can be seen in the photomicrograph of a cross section of several particles in Fig. 1a.

A second kind of polymer, a colloidal aqueous dispersion, was reported by RENFREW (1950) who used bis- (β-carboxypropionyl) peroxide as the polymerization initiator, and later described in more detail by LONTZ and HAPPOLDT. The specific surface of dispersion polymer is on the order of 12 m^2/g, and the equivalent surface average diameter for dense spheres is about 0.2 μ. This is a good check with the observed size seen in the electron micrograph of Fig. 1b and indicates that the primary dispersion particles have little, if any, porous structure.

HANFORD and JOYCE placed great emphasis on the precautions which are required for the safe handling of tetrafluoroethylene, particularly at elevated pressures. Failure to provide for the adequate control of temperature and efficient agitation may lead to an increasingly rapid reaction and eventually a violent explosive decomposition of the monomer to carbon and CF_4 (KIYAMA, OSUGI, and KUSUHARA; TERANISHI).

III. Molecular Structure

A. Chain Structure

Polytetrafluoroethylene is believed to be an entirely linear polymer (McGREW). The evidence for the linear structure rests on the high level of crystallinity as polymerized, the infrared spectra, and energetic arguments against the likelihood of chain transfer reactions such as those which lead to branching in polyethylene (ROEDEL). Specifically, the density of polytetrafluoroethylene at room temperature is about 2.0 and 2.3 g/cm^3 for the amorphous and crystalline portions, respectively. In most polymerizations of tetrafluoroethylene, the solid polymer (rather than a melt) is formed directly from the monomer. The density of this polymer has been found to be between 2.28 and 2.295 g/cm^3 which corresponds to a degree of crystallinity of 93—98%. Experience with other polymers indicates that such a high level of crystallinity could be obtained only when the chain structure is substantially unbranched.

In polyethylene, branching occurs by chain transfer to polymer during the polymerization (ROEDEL). An analogous reaction in polytetrafluoroethylene would require the breaking of a CF bond which has a dissociation energy of about 110 kcal per mole. Since this is one of the strongest bonds in organic compounds, it has been argued on the basis of kinetic and thermodynamic factors that this reaction is very unlikely under the usual conditions for polymerization (BRYANT).

B. End Groups

The first report on the nature of the end groups of polytetrafluoroethylene was made by BERRY and PETERSON, who studied the polymerization of tetrafluoroethylene with initiator systems employing persulfate or bisulfite containing radio-active sulfur, S^{35}*. They showed that the sulfur from the persulfate initiator does not appear in the polymer. In contrast, however, polymers made with bisulfite do possess end groups containing sulfur that were not hydrolyzed during the polymerization. The following equations show how bisulfite can enter the polymer to form sulfonic acid end groups.

Initiation:

$$Fe^{+++} + HSO_3^- \rightarrow Fe^{++} + HSO_3\cdot$$

Propagation:

$$HSO_3\cdot + n(CF_2{=}CF_2) \rightarrow HO\overset{\overset{O}{\uparrow}}{\underset{\underset{O}{\downarrow}}{S}}(CF_2CF_2)_n\cdot$$

$$HSO_3\cdot + m(CF_2{=}CF_2) \rightarrow HO\overset{\overset{O}{\uparrow}}{\underset{\underset{O}{\downarrow}}{S}}(CF_2CF_2)_m\cdot$$

Termination:

$$HO\overset{\overset{O}{\uparrow}}{\underset{\underset{O}{\downarrow}}{S}}(CF_2CF_2)_n\cdot + HO\overset{\overset{O}{\uparrow}}{\underset{\underset{O}{\downarrow}}{S}}(CF_2CF_2)_m\cdot \rightarrow HO\overset{\overset{O}{\uparrow}}{\underset{\underset{O}{\downarrow}}{S}}(CF_2CF_2)_{(n+m)}\overset{\overset{O}{\uparrow}}{\underset{\underset{O}{\downarrow}}{S}}OH$$

Since sulfur from the persulfate does not appear in the polymer, two possible reactions were suggested. The sulfate ion-radicals may initiate the polymerization to produce fluoroalkyl sulfuric acid esters which would be very rapidly hydrolyzed.

$$\left[O\overset{\overset{O}{\uparrow}}{\underset{\underset{O}{\downarrow}}{S}}OO\overset{\overset{O}{\uparrow}}{\underset{\underset{O}{\downarrow}}{S}}O\right]^{=} \rightarrow 2\left[O\overset{\overset{O}{\uparrow}}{\underset{\underset{O}{\downarrow}}{S}}O\right]^{\dot{-}}$$

$$\left[O\overset{\overset{O}{\uparrow}}{\underset{\underset{O}{\downarrow}}{S}}O\right]^{\dot{-}} + n(CF_2{=}CF_2) \rightarrow O\overset{\overset{O}{\uparrow}}{\underset{\underset{O}{\downarrow}}{S}}O(CF_2CF_2)_{\overline{n}} \xrightarrow{H_2O} [HO(CF_2CF_2)_{\overline{n}}]$$

Alternatively, the sulfate ion-radicals may hydrolyze to give hydroxyl radicals which could initiate the polymerization.

$$\left[O\overset{\overset{O}{\uparrow}}{\underset{\underset{O}{\downarrow}}{S}}O\right]^{\dot{-}} + H_2O \rightarrow HSO_4^- + HO\cdot$$

$$HO\cdot + n(CF_2{=}CF_2) \rightarrow [HO(CF_2CF_2)_{\overline{n}}]$$

In either case difluorocarbinol end groups would result and these would be rapidly hydrolyzed to carboxyl groups.

$$HO(CF_2CF_2)_{\overline{n}} \xrightarrow{H_2O} HO\overset{\overset{O}{\|}}{C}CF_2(CF_2CF_2)_{\overline{n-1}} + 2\,HF$$

In later work, carboxyl groups were identified in polytetrafluoroethylene initiated with persulfate (BRO and SPERATI). Using infrared techniques and comparative model compounds, it was shown that the carboxyl end groups could be converted to the sodium salt and then pyrolyzed to give perfluorovinyl groups.

$$\mathrm{HO\overset{\overset{\displaystyle O}{\|}}{C}(CF_2{-}CF_2)_{\overline{n}}} \xrightarrow{\text{NaOH}} \mathrm{NaO\overset{\overset{\displaystyle O}{\|}}{C}(CF_2CF_2)_{\overline{n}}} \xrightarrow{\Delta} \mathrm{CF_2{=}CF(CF_2CF_2)_{\overline{n-1}}}$$

BRYANT has calculated the changes in free energy for various reaction steps of the polymerization of tetrafluoroethylene. He concluded (1) that the initiation and propagation are about twice as favorable for tetrafluoroethylene as the analogous reactions for ethylene, (2) that termination by combination is more favorable than disproportionation, and (3) that chain-transfer to monomer and to polymer are less likely than the combination of radicals.

Since reactions of free radicals with saturated molecules have appreciable energies of activation and also have negative entropies of activation, termination by disproportionation and transfer to monomer and polymer may be even less favorable than the changes in free energy would indicate. Moreover, BRO and SPERATI pointed out that since the characteristic absorption of perfluorovinyl groups at 5.60μ in the infrared spectrum was not observed in virgin polymers, termination by disproportionation does not occur to any great extent. Thus, it is considered highly probable, although not definitely proved, that each molecule of polytetrafluoroethylene contains two sulfonic or carboxyl end groups depending on the type of initiator used.

C. Molecular Weight

Polytetrafluoroethylene is insoluble at moderate temperatures, and the usual methods for measuring molecular weight, which employ dilute solutions, are not applicable. The only quantitative estimates of the molecular weight have been based on determinations of the concentration of end groups derived from the initiator. The validity of this procedure, which gives a number-average molecular weight, depends on the assumptions outlined in the preceding section. Early measurements of molecular weight were made by BERRY and PETERSON, who used the iron-bisulfite system containing radioactive sulfur. They reported number-average molecular weights from 142,000 to 534,000 on specially prepared low-molecular-weight polytetrafluoroethylenes. Subsequently, DOBAN, KNIGHT, PETERSON, and SPERATI applied similar techniques to polymers having properties similar to those of industrial interest. Their samples varied between 389,000 and 8,900,000 in number-average molecular weight.

In practice, it is customary to estimate the relative molecular weight by measuring the specific gravity following a standard fabricating cycle. The test for this standard specific gravity (SSG) is part of the ASTM specification for tetrafluoroethylene resin molding and extrusion materials designated D 1457-56T. Since the rate of crystallization varies inversely with molecular weight, samples of high molecular weight have a lower standard specific gravity than those of low molecular weight. A problem in use of this procedure is the presence of voids in some samples which results in a low value for the measured specific gravity. It is possible to correct for this, however, by applying infrared techniques which are described in a later section.

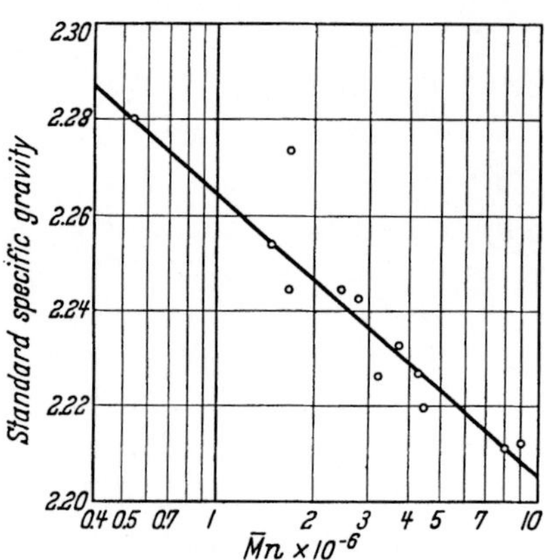

Fig. 2. Dependence of the standard specific gravity on the molecular weight calculated from the concentration of end groups tagged with radioactive sulfur. (According to DOBAN, KNIGHT, PETERSON and SPERATI)

The dependence of the standard specific gravity on the number-average molecular weight is shown in Fig. 2.

IV. Crystalline Structure

A. X-Ray Diffraction

The first published study of the diffraction of x-rays by the crystal structure of polytetrafluoroethylene was by BUNN and HOWELLS. Later, PIERCE, CLARK, WHITNEY, and BRYANT described the crystal structure

ABOVE 19°C

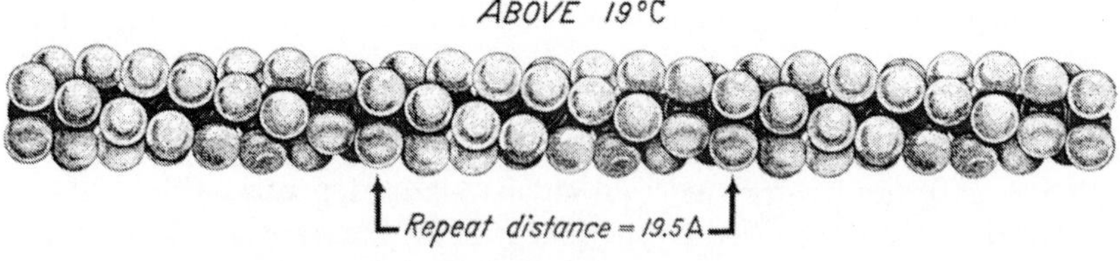

BELOW 19°C

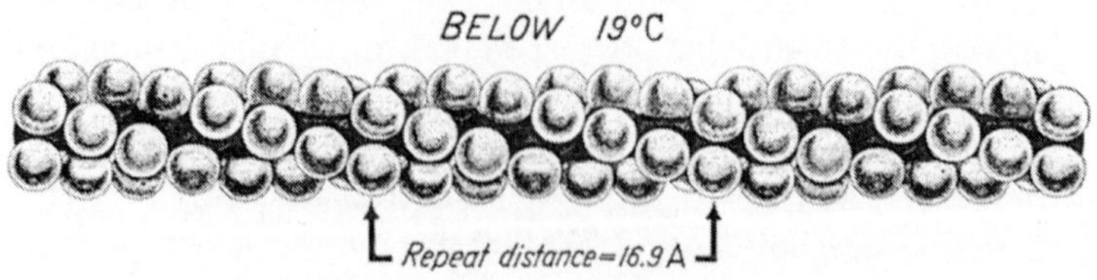

Fig. 3. Models of the polytetrafluoroethylene chain above and below the crystalline transition at 19° C

in terms of the conformation of the chain molecules, that is, the geometrical arrangement of the atoms, and the nature of the packing of the molecules to form the crystal. The chain conformation was treated mathematically in terms of atoms regularly spaced along helices, an approach equivalent to defining the contents of the unit cell in terms of cylindrical coordinates. Packing of the chains was characterized by defining the distances between molecules as determined from the equatorial reflections on an x-ray fiber pattern.

The crystal structure of polytetrafluoroethylene shows two sharp changes at about 19° C and 30°C (PIERCE, CLARK, WHITNEY, and BRYANT). These changes are generally considered to be first-order crystalline transitions. The molecular conformation below and above the 19° C transition are shown in drawings of molecular models in Fig. 3 and the changes in the diffraction patterns of crystalline fibers at these transitions are illustrated in Fig. 4. Below 19° C the

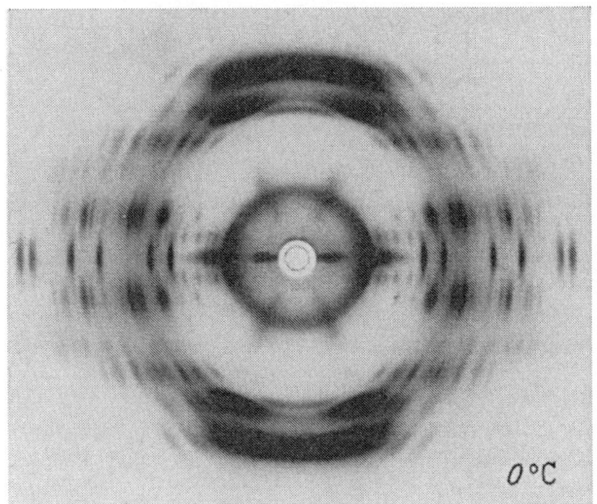

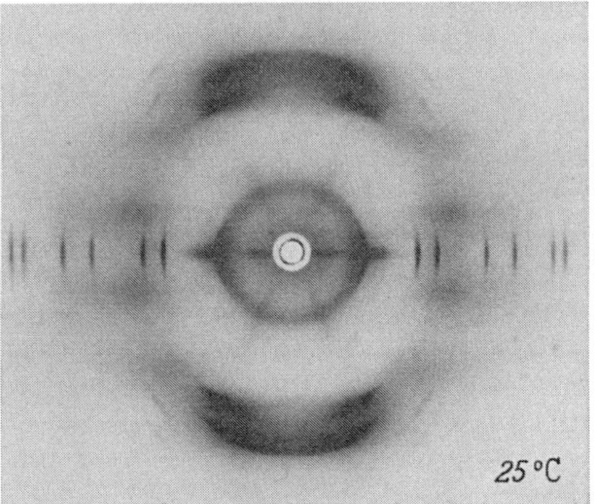

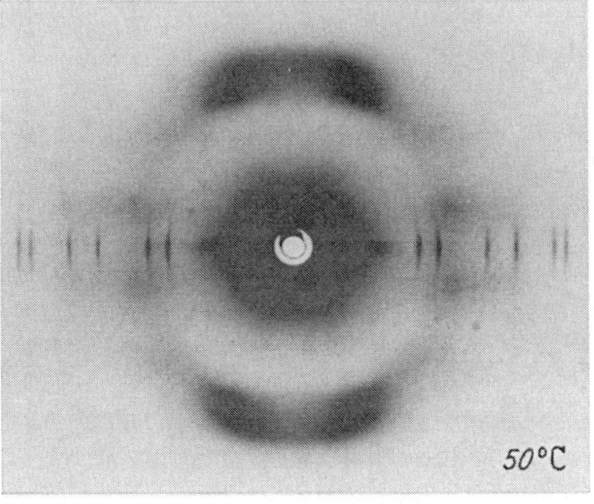

Fig. 4. X-Ray diffraction patterns of oriented fibers above and below the crystalline transition at 19° and 30°C

conformation is that of a zig-zag arrangement which has been given a twist of 180° per 13 carbon atoms. The repeat unit contains 13 CF_2 groups with a repeat distance of 16.9 Å. The x-ray fiber pattern shows two intense upper layer lines, the sixth and seventh layers, while the other upper layer lines are either weak or missing. The large number of reflections on the upper layer lines shows that the unit cell is skewed in shape, probably triclinic. On the other hand, the few reflections on the zero layer line show that, in the lateral direction, the molecules pack like cylindrical rods in a nearly hexagonal arrangement with a nearest neighbor distance of 5.62 Å.

Above 19° C, the triclinic pattern changes to a new one with layer lines showing both diffuse streaks and sharp spots characteristic of a hexagonal unit cell. The distribution of intensity in the fiber pattern shows that the molecules retain a twisted zig-zag conformation. The two intense upper layer lines are indexed as the seventh and eight layers. This assignment is in contrast to the low-temperature form for which the sixth and seventh layer lines are intense. This difference results from a slight untwisting of the molecule in the 19° transition from a 180° twist per 13 CF_2 groups to an 180° twist per 15 CF_2 groups. At 25° C, the repeat distance is 19.5 Å corresponding to the repeat unit of 15 CF_2 groups and the separation of the chain axes is 5.66 Å. The diffuse streaks in the layer lines imply disorder in the packing of the molecules in the lattice.

The sharp spots on the upper layers that are indicative of a hexagonal unit cell disappear above the 30° transition. As shown by the sharp equatorial spots, the rod-like hexagonal packing of the chains in the lateral direction is retained. Diffraction patterns taken at still higher temperatures suggest an irregular twist in the molecule, but the lateral packing of the chains remains hexagonal to the melting point of 327° C.

CLARK and MUUS (1957, 1958) subsequently made more detailed studies of the disorder in the crystal lattice of polytetrafluoroethylene using helical structure concepts. Because of the ordered hexagonal lateral packing of the chains, disorder was limited to five types: (1) small translations along the chain axis, (2) large translations along the chain axis, (3) small angular displacements about the chain axis, (4) large angular displacements about the chain axis, and (5) screw displacements along the chain axis. Mathematical expressions were derived for the effect of each type of disorder on the diffraction pattern. Comparison of these equations with experimental measurement showed that below the 19° C transition there is essentially perfect three-dimensional order. Between 19° C and 30° C, the chain segments are disordered from a perfect hexagonal lattice by small angular displacements about their longitudinal axes.

Since x-ray diffraction presents a time-space average of the crystal structure, it cannot be determined directly whether the segments are oscillating or are statistically distributed over a range of orientations. Theoretical considerations, however, favor small-angle oscillations of molecular segments about their long axes with the angular displacements symmetrically distributed about a preferred crystallographic direction. Above the 30° C transition, Bragg peaks are observed only on the equator and the 15th layer; no spots appear on the intervening layers. It was concluded that the preferred crystallographic direction has been lost and that the molecular segments oscillate about their long axes with a random angular orientation in the lattice (CLARK, 1959). Changes in the x-ray patterns at still higher temperatures suggest torsional motion, a twisting and untwisting of the chain molecules.

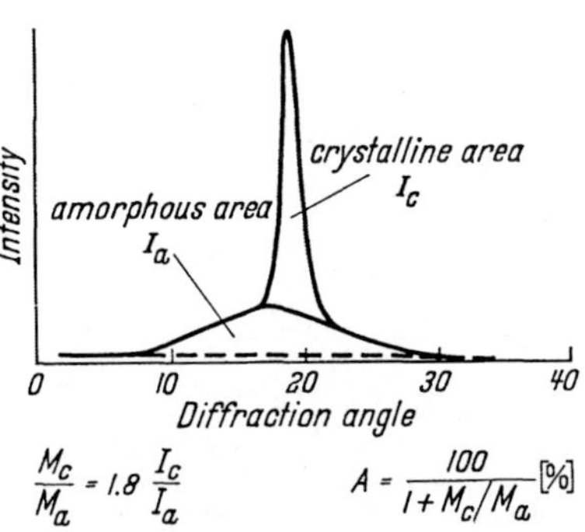

Fig. 5. Calculation of the amorphous content of polytetrafluoroethylene from the x-ray diffraction pattern. (After RYLAND)

In diffraction patterns made from unoriented samples, the crystalline pattern is superimposed on an amorphous halo. As shown in Fig. 5, the percentage of amorphous material may be calculated by comparing the intensities of the two portions of the diffraction pattern. When the amorphous fraction is large, as in samples which have been quenched rapidly from the melt to a low temperature, the crystal structure may be greatly disrupted. Various interpretations of the intermolecular and intramolecular order which may be deduced from the diffraction patterns of such samples are discussed by KILIAN and JENCKEL.

B. Infrared Spectra

While the application of the methods of group theory to the analysis of spectra from high polymers might seem hopelessly complex, the great length of polymer molecules and the assumption of well-defined order in the crystalline state actually contribute to a considerable simplication of the problem. Band assignments for the spectrum of the polytetrafluoroethylene crystal have been made on this basis by LIANG and KRIMM and by MOYNIHAN. In order to make use of the crystal symmetry in calculating the spectra when experimental samples are only partially crystalline, it is necessary to determine which bands vary with the degree of crystallinity and then extrapolate the spectra to 100% crystallinity. The most pronounced changes in the spectra with respect to crystallinity occur in the region of 700—850 cm^{-1} (11.7—14,3 μ) region and at 384 cm^{-1}

(26.0 μ). In these regions, several bands were found which decrease regularly with increasing crystallinity (increasing density). These bands are obviously characteristic of the amorphous portions of the polymer. Some features of the spectrum are clearly related to the crystalline order. The absorption at 625 cm^{-1} (16.0 μ), for example, is present only above the crystal-disordering transition at 19° C. This band has been assigned to a CF_2 wagging mode.

MOYNIHAN made particular use of the band at 778 cm^{-1} (12.85 μ) for which the absorbance per unit thickness divided by density is proportional to the amorphous content measured by x-ray diffraction. In Fig. 6, it is seen that the absorbance per unit thickness for this band is a linear function of the density. This curve extrapolates to zero absorbance for a density of 2.304 ± 0.006 g/cm^3 at 23° C, which is in excellent agreement with the crystalline density of 2.301 g/cm^3 calculated from the measurements of x-ray diffraction by CLARK and MUUS. The amorphous density however, must be calculated by extrapolating x-ray data or by the comparison of liquid fluorocarbons of low molecular weight. The most probable value is 2.00 ± 0.04 g/cm^3.

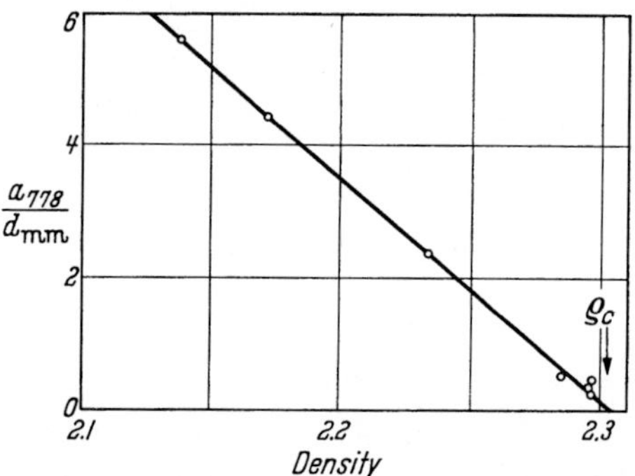

Fig. 6. Extrapolated crystalline density for polytetrafluoroethylene. (After MOYNIHAN). ϱ_c = Crystalline density calculated from the unit cell determined by x-ray diffraction

A routine method for determining relative crystallinity based on the amorphous bands in the spectrum has proved more rapid and precise than the x-ray method. In practice, the ratio of the 778 cm^{-1} (12.85 μ) and 2367 cm^{-1} (4.22 μ) band intensities is measured. Use of a ratio eliminates the thickness measurement and increases precision to about ± 1% at 50% crystallinity and considerably better at higher levels. A density measurement and an infrared crystallinity determination when combined give an estimate of the fraction of microvoids which can occur in molded specimens of polytetrafluoroethylene. The density of a sample is predicted on the basis of its crystallinity as measured by the infrared method and the difference between this density and the actual density measured by displacement in water is a measure of the microvoid content. This determination is precise to about ± 0,2% voids by volume. By the use of confirmatory infrared measurements, it is possible to check the possibility that the presence of a substantial percentage of voids may have led to erroneous indications of the molecular weight in the standard specific gravity test discussed earlier.

C. Microscopy

Studies of the fine structure of polytetrafluoroethylene using electron and light microscopes have been reported by BUNN, COBBOLD, and PALMER. Granular polymer was molded at 380° C, cooled slowly, and then fractured in liquid nitrogen. The fracture surfaces contain long bands from 0.2 μ to 1 μ thick. Perpendicular to these long bands are striations varying down to 300 Å in thickness. From the birefringence of the samples, it was determined that the polymer chains are parallel to the striations and perpendicular to the long bands. The sharpness of the spots in the x-ray diffraction pattern indicates that the crystallites are at least 1,000 Å thick. When the polymer was treated at 500° C and then cooled, the bands were somewhat disordered to give structures approaching the familiar spherulites of other polymers. From these observations, it was inferred that a considerable amount of structure persists above the melting point.

While this well defined band structure was originally interpreted as indicating a very narrow molecular-weight distribution, it now seems more probable to the present authors that the chains are folded as they are in other crystalline polymers. The long fold period may well be related to the great stiffness of the fluorocarbon chains.

D. Thermodynamic Properties

1. Heat Capacity at Very Low Temperatures. The heat capacity of polytetrafluoroethylene from 0° to 365° K was measured by FURUKAWA, McCOSKEY, and KING. They found that samples with different thermal histories all have nearly the same heat capacity up to about 160° K (—113° C). At somewhat higher temperatures the heat capacity is slightly higher for samples having lower levels of crystallinity. This indicated that a glass transition might occur at about 160° K. The heat capacity per carbon atom is considerably higher for polytetrafluoroethylene than for polyethylene. This is because the greater mass of the CF_2 group compared with the CH_2 group reduces the characteristic lattice frequencies and thereby increases the amount of thermal motion at low temperatures (STARKWEATHER). In addition, vibrations of the CF bonds are excited at much lower temperatures than those of CH bonds. At temperatures as low as 80° K, for example, the heat capacity is larger than would be expected for rigid CF_2 groups. The ductility of polytetrafluoroethylene at very low temperatures (SWENSON, 1954; DYMENT and ZIEBLAND) is probably related to the fact that there is more thermal motion than in hydrocarbons, making it easier for the chains to slip past each other.

In a fundamental approach to calculation of the properties of high polymers, BRANDT has estimated the lattice energy and the compressibility of polytetrafluoroethylene at absolute zero by first developing the intermolecular potential function.

2. Crystal-Disordering Transitions. The evidence for structural changes at 19° and 30° C contained in x-ray diffraction patterns has been discussed in the preceeding section. (We will continue to refer to the transitions by these temperatures even though they are observed at slightly different temperatures in some samples). These crystalline transitions were first discussed in a report on thermal expansion by RIGBY and BUNN and were confirmed in subsequent studies using dilatometry (QUINN, ROBERTS, and WORK; LEKSINA and NOVIKOVA) and calorimetry (FURUKAWA, MCCOSKEY, and KING; MARX and DOLE).

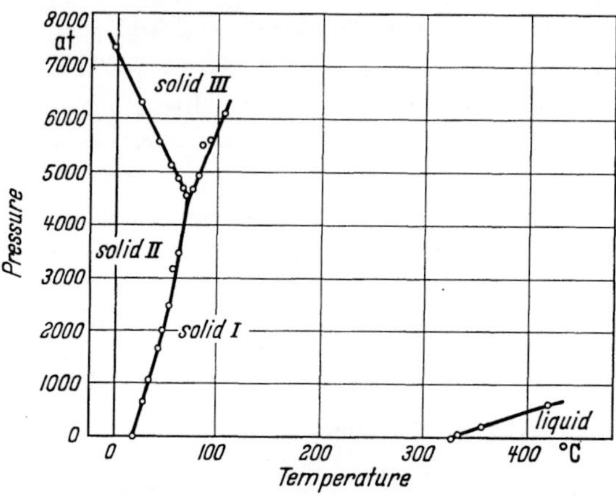

Fig. 7. Phase diagram for polytetrafluoroethylene. (Low temperature transitions according to BEECROFT and SWENSON, melting according to MCGEER and DUUS)

The transition at 19° C involves an expansion of 0.0058 cm³/g (CLARK and MUUS). Since the transition temperature increases with pressure by about 0.013° C per atmosphere (BEECROFT and SWENSON), the latent heat is about 3.2 cal/g. These values are for the crystal and would be reduced in proportion to the crystalline content. The transition at 30° C is only about one-tenth as large. The over-all increase in entropy at these transitions is about 0.0108 cal $deg^{-1} g^{-1}$. The portion due to the increase in volume is $(\alpha/\beta)\Delta V$, where α is the volumetric coefficient of thermal expansion and β is the compressibility. Since the compressibility of the crystal is not known, this quantity is somewhat uncertain. Using the average of the values of α (QUINN, ROBERTS, and WORK) and β (WEIR, 1951) for the whole polymer above and below the transitions, it appears that $(\alpha/\beta)\Delta V$ is about 0.0041 cal $deg^{-1} g^{-1}$. The entropy of the transition corrected to constant volume is, therefore, about 0.0067 cal $deg^{-1} g^{-1}$.

MARX and DOLE and MIYAKE have presented descriptive models for the 19° C transition in terms of order-disorder theories. Studies of transitions at high pressure in polytetrafluoroethylene have been reported by BRIDGMAN, WEIR (1953), and BEECROFT and SWENSON. The phase diagram in Fig. 7 shows that in addition to the two crystalline phases which are separated by the 19° transition at atmospheric pressure there is a third modification at high pressures. The triple point has been located at 70° C and 4,500 atmospheres. There is also evidence for another transition above room temperature and at pressures near 11,000 atmospheres.

3. Melting. Polytetrafluoroethylene melts at 327° C with a heat of melting about 685 cal mole^{-1} of CF_2 groups (LUPTON). Thus, the entropy of melting is 1.14 cal deg^{-1} mole^{-1} of carbon atoms vs 2.29 for polyethylene. It has been shown by STARKWEATHER and BOYD that 0.38 cal deg^{-1} mole^{-1} of CF_2 groups of this increase in entropy is due to the expansion of about 29% which occurs when the crystal melts. Thus, the entropy of melting corrected to constant volume is 0.76 cal deg^{-1} mole^{-1} of carbon atoms vs 1.8 for polyethylene. The high melting point of polytetrafluoroethylene is due to the low entropy of fusion which in turn results from the high stiffness or persistence of direction of the polymer chains. The stiffness is also reflected in the x-ray diffractometer scan for molten polytetrafluoroethylene reported by KILIAN and JENCKEL. There are two intensity maxima, one corresponding to the average separation between the chains and the other to the repeat distance along the chain. The presence of the second maximum means that the chains are almost straight for several repeat units.

An increase in the melting point with increasing pressure (0.154 deg per atmosphere) was reported by McGEER and DUUS, and the pressure-volume-temperature relationship of polytetrafluoroethylene above its melting point was studied in more detail by LUPTON who found the following equation of state:

$$(P + a)\,(V - b) = B\,(T - c)$$

where: a = 400 atmospheres
b = 0.500 cm^3/g
c = 144° C
B = 0.306 cm^3-atm g^{-1}deg^{-1}

The specific volume was slightly lower for a sample of very high molecular weight and was reflected by an increase in c to 156° C. This effect was attributed to the very high melt elasticity of this polymer. One consequence of the finite value of c is that the internal pressure is not constant as it is for melted polyethylene but decreases with increasing volume.

V. Mechanical Properties

A. Viscoelastic Behavior

In a study of internal friction in polytetrafluoroethylene, McCRUM (1959a) investigated samples varying from 48% to 92% crystallinity. The dependence of the logarithmic decrement and the torsion modulus on temperature is shown in Fig. 8 and 9. There are four relaxation regions in the temperature range from 4.2 to 600° K (—269 to 327° C). The relaxations are classified as being either in the amorphous or the

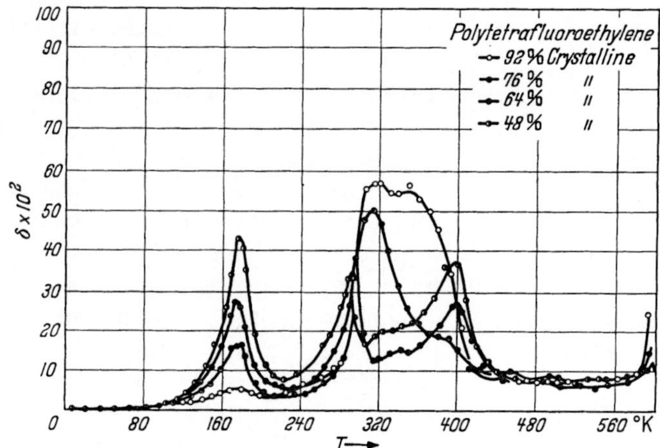

Fig. 8. The internal friction of four samples of polytetrafluoroethylene of various crystallinities from 4.2 to 600° K. (After McCRUM)

crystalline portion of the polymer, depending on (1) the variations with crystallinity of the height of the damping maximum, (2) the magnitude of the corresponding change in the modulus, or (3) evidence derived from

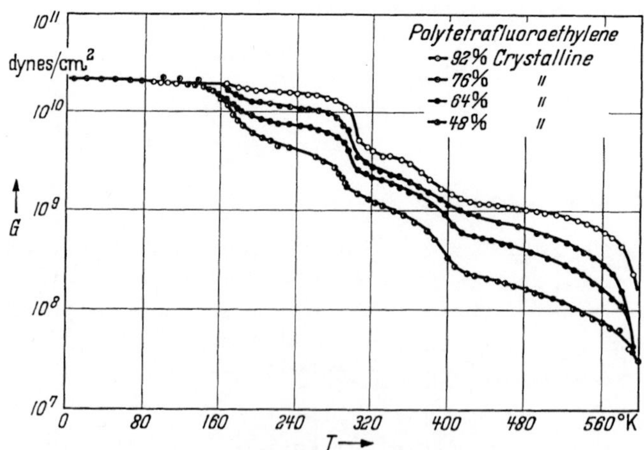

Fig. 9. The torsion modulus of four samples of polytetrafluoroethylene of various crystallinities from 4.2 to 600° K. (After McCRUM)

other experiments, e. g., x-ray diffraction, specific heat. Because the heights of the loss maxima at the —97° C and 127° C transitions increase with decreasing crystallinity, these relaxations must occur in the amorphous regions of the polymer. These transitions have been designated Glass II and Glass I, respectively. In the temperature region of the crystal-disordering transition, near 20° C, the relaxation effects increase with increasing crystallinity and occur in the crystalline portions of the polymer. Finally, there is a small relaxation just below 327° C which is undoubtedly due to the melting of the crystallites.

The amorphous transition at —97° C was first detected in measurements of the specific heat. (FURUKAWA, McCOSKEY, and KING). It is attributed to motions involving small sequences of CF_2 groups. The dependence of the temperature of this transition on the frequency of oscillation from 0.15 cps to 5 × 10^6 cps is shown in Fig. 10 (McCRUM, 1959b).

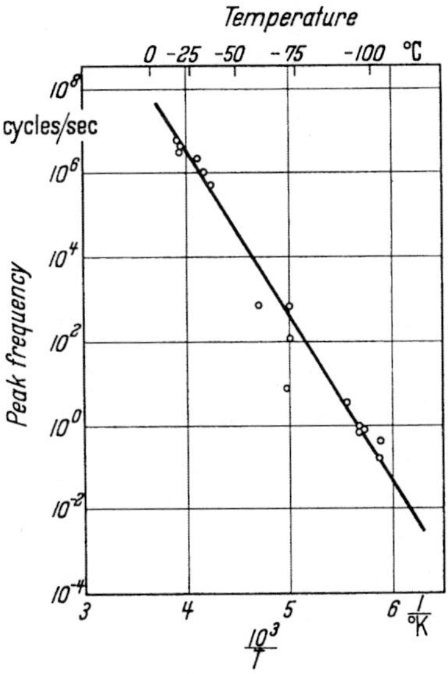

The transition has also been detected in dielectric studies at frequencies from 10^3 to 10^7 cps (MIKHAILOV, KABIN, and SMOLYANSKII). The activation energy for the transition is 18 kcal per mole, and the activation entropy is 45 entropy units per mole. From the breadth of the loss peak in Fig. 8, it has been determined that more than a single relaxation time is required.

Fig. 10. The dependence of the glass II internal friction peak on temperature. (After McCRUM)

The amorphous transition at 127° C is depressed to lower temperatures in copolymers of tetrafluoroethylene with up to 14 mole per cent hexafluoropropylene (McCRUM, 1959b). For this reason it is attributed to longer sequences of CF_2 groups.

ILLERS and JENCKEL have discussed the internal frictions peak in the region of the crystal disordering transition. They analyzed the shape of the loss peak at various frequencies in terms of two different distributions of relaxation times existing above and below the transition. NAGAMATSU, YOSHITOMI, and TAKEMOTO have also discussed this change in the distribution of relaxation times as revealed by measurements of stress relaxation.

Nuclear magnetic resonance provides a valuable complement to internal friction in studies of molecular motion on polytetrafluoroethylene. Pertinent studies have been reported by SMITH, SLICHTER, WILSON and PAKE, POWLES and KAIL, and by SINNOTT. HYNDMAN and ORIGLIO studied the NMR spectra of drawn fibers. They found that the anisotropy is greater above 12° C. The spectra suggest that above the crystal disordering transition, chain segments undergo occasional 360° rotations about the chain axis. This would be in addition to the torsional oscillations indicated in studies of x-ray diffraction.

B. Strength and Stiffness

The physical properties of polytetrafluoroethylene have been reviewed by DOBAN, SPERATI, and SANDT and by RENFREW and LEWIS, and some average properties for well molded samples are listed in Table 1. The effects of variations in the degree of crystallinity and the molecular weight have been discussed by SPERATI and MCPHERSON and related to the details of the fabrication procedures by THOMAS, LONTZ, SPERATI, and MCPHERSON. Crystalline contents in most polymers can be varied over a wide range only by changing the molecular structure, e.g., by varying the amount of chain branching in polyethylene. In polytetrafluoroethylene, on the other hand, it is possible to obtain a wide variation in the degree of crystallinity by varying the rate at which the polymer is cooled from the melt. The crystalline content has a marked effect on some properties and Fig. 11, for example, shows that the flexural modulus is extremely sensitive to the per cent crystallinity. The analogous relationship for the torsion modulus was used by MCCRUM (1959c) to develop a method for measuring the degree of crystallinity and the void content in moldings. Not all of the physical properties are improved by increasing the per cent crystallinity. The tensile impact strength drops about eight-fold as the crystallinity is increased from 60 to 90 %. Some properties go through maxima or minima with increasing crystallinity. One example is the apparent modulus after 100 hours' creep under a stress of 1,000 lb/sq in. This property shows a sharp maximum at 75 to 80% crystallinity. This type of behavior was explained in the following

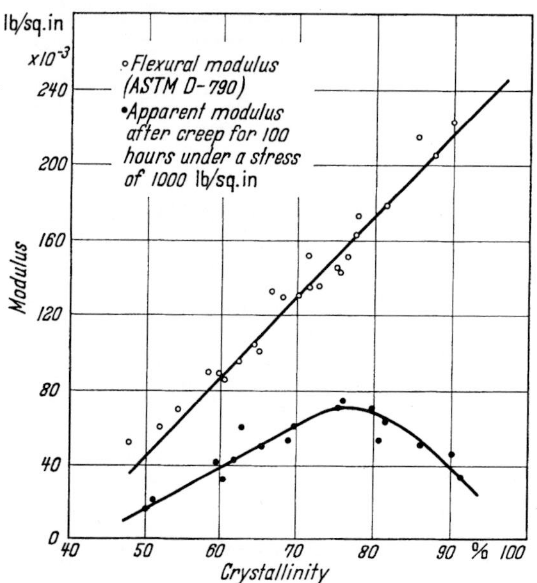

Fig. 11. Dependence of the modulus on per cent crystallinity (According to SPERATI and MCPHERSON)

way: Increasing the crystalline content is analogous to increasing the number of crosslinks in a rubber and thereby increasing the resistance to deformation. For high levels of crystallinity, however, stresses tend to become concentrated at certain points causing increased slip and a lower apparent modulus after creep.

Table 1. *Typical Physical Properties of Polytetrafluoroethylene*

Property	Units	Method
Tensile strength, 23° C	2500–4000 lb/sq in	D 638-58T
Elongation, 23° C	200–400%	D 638-58T
Flexural strength, 23° C	did not break	D 790-59T
Stiffness, 23° C.	60,000 lb/sq in.	D 747-58T
Impact strength, Izod		
−57° C	2.0 ft.-lb/in.	D 256-56
23° C	3.5 ft.-lb/in	D 256-56
77° C	6.0 ft.-lb/in	D 256-56
Hardness, durometer	D 55–D 70	D 1706-59T
Compressive stress at 1% deformation, 23° C	600 lb/sq in	D 695-54
at 1% offset, 23° C	1000 lb/sq in	D 695-54
Deformation under load, 50° C		
1200 lb/sq in., 24 hrs	4–8%	D 621-59
2000 lb/sq in., 24 hrs	25%	D 621-59
Deflection temperature under load		
66 lb/sq in	250° F	D 648-56
Coefficient of linear thermal expansion		
per °F, 25–60° C	5.5×10^{-5}	D 696-44
Thermal conductivity, 0.18 in	2.1 cal/hr/sq cm °C/cm	(1)
Specific heat	0.25 cal/g/°C	
Water absorption	0.0%	D 570-59T
Flammability	Nonburning	D 635-56T
Specific gravity	2.1–2.3	D 792-50
Resistance to weathering	Excellent (2)	

Tests have been performed by ASTM methods unless otherwise indicated. Data shown are average values and should not be used for specifications.

(1) Thermal conductivity measured by Cenco-Fitch apparatus.

(2) No detectable change after 10 years of outdoor exposure in Florida.

Samples of polytetrafluoroethylene having very high degrees of crystallinity are characterized by a very low yield strain and exhibit lead-like behavior after deformation. In contrast with other crystalline polymers the yield stress does not increase with increasing crystallinity, but remains almost constant at room temperature. At lower temperatures the yield stress decreases with increasing crystallinity (Riley).

Above a relatively low molecular weight of about one-half million, the physical properties which are discussed above depend only on the per cent crystallinity. Other properties are strongly dependent on the molecular weight and void content (Thomas, Lontz, Sperati, and McPherson).

The maximum changes in properties which can be produced by variations in molecular weight, per cent crystallinity, and void content are summarized in Table 2.

Table 2. *Effect of Molecular Weight, Crystallinity, and Void Content on Properties*

Property	Maximum Change Due to Increase in		
	Mol. Wt.	Crystall.	Void Content
Flex Fatigue Life	+100 fold	−100 fold	−1000 fold
Compressive Stress at 1% Deformation	0	+50%	0
Compressibility	0	−50%	No Data
Recovery	0	−70%	No Data
Permeability to CO_2	0	−30 fold	+1000 fold
Flexural Modulus	0	+ 5 fold	−30%
Hardness-Durometer	0	+20%	No Data
Hardness-Rockwell	0	−20%	−30%
Hardness-Scleroscope	0	−70%	−10%
Tensile Impact Strength	0	−15 fold	−80%
Dielectric Strength	0	0	−70%
Proportional Limit	0	+80%	−20%
Yield Stress	0	+15%	−20%
Yield Strain	0	−15 fold	0
Tensile Strength	+25%	−50%	−50%
Ultimate Strength	+50%	−70%	−50%
Ultimate Elongation	−20%	+100%[1]	−80%

[1] Reaches a maximum at 85% crystallinity. Drops to $^1/_{10}$ maximum above 85% crystallinity.

C. Surface Properties

Early in the history of polytetrafluoroethylene, W. A. Zisman recognized its unusual surface properties, and the polymer now finds many uses because its low coefficient of friction eliminates the need for lubrication (FITZSIMMONS and ZISMAN). SHOOTER and THOMAS called attention to the remarkable resistance of polytetrafluoroethylene to seizure, and BOWDEN (1950) described use of composite structures. These and other investigations have been reviewed and summarized by ALLAN (1958) (also ALLAN and CHAPMAN), who showed that the dependence of the coefficient of kinetic friction, f_k, on the load in grams, W, is given by the following equation:

$$f_k = 0.178\, W^{-0.5}$$

The coefficient of friction increases with increasing sliding speed up to about 150 ft/min, while beyond this point, the sliding speed has very little effect. Polytetrafluoroethylene filled with molybdenum disulfide has a coefficient of friction lower than for pure polymer under the same conditions. The decrease is attributed to the increased hardness of the mixture as well as to the flake-like structure of molybdenum disulfide. The coefficient of friction is relatively independent of temperature above room temperature, although some samples show an abrupt drop on cooling through the 19° C transition.

Differences in the frictional properties of most plastics can be explained in terms of the ratio of shear strenghth to hardness. SHOOTER and TABOR observed that the coefficients of friction for polytetrafluoroethylene are 2—3 times lower than anticipated by this calculation. It is believed that this discrepancy is caused by the inherently low cohesive forces between adjacent polymer chains and is responsible for the absence of stick-slip. The large fluorine atoms effectively screen the large carbon-fluorine dipole, reducing molecular cohesion so that the shear force at the interface is low. The shear strength of the bulk material is higher because of interlocking molecular chains.

FOX and ZISMAN studied the spreading of various liquids on surfaces of polytetrafluoroethylene, and found a linear increase in the cosine of the contact angle as the surface tension of the liquid decreased. Liquids having surface tensions below about 20 dynes/cm spread completely. Thus, solutions of various perfluorocarbon acids in water will wet the polymer (BERNETT and ZISMAN). ALLAN and ROBERTS studied the differences between the advancing and receding contact angles formed by a drop of water. The hysteresis increased with increasing surface roughness. This is important when surfaces of polytetrafluoroethylene are employed to prevent the adherence of ice.

Surface treatments involving alkali metals are sometimes used to eliminate the characteristic surface properties and promote the adhesion between polytetrafluoroethylene and other substances (DOBAN; NELSON, KILDUFF, and BENDERLY; PURVIS and BECK; RAPPAPORT). It has been shown that these treatments produce a marked increase in the polarity of the surface as measured by the contact angle with various liquids (ALLAN, 1957). They also increase the coefficient of friction. One interesting application of surface properties of polytetrafluoroethylene was reported by BOWDEN (1953, 1955) who applied the polymer to the bottoms of his skiis and thereby reduced the friction between the skiis and the snow.

VI. Electrical Properties

Many important applications of polytetrafluoroethylene depend on its superb electrical properties tabulated in Table 3. These properties have been attributed to its highly symmetrical structure (DOBAN, SPERATI, and SANDT). Complete fluorination of the carbon chain results in an exact balance of the electrical dipoles which is manifested in a very low dielectric constant and electrical loss factor. These two properties are virtually independent of the frequency from 60 to 10^9 cycles per second

(VON HIPPEL), and over a frequency range of 10^2 to 10^5 cycles per second, the dissipation factor does not change within the limits of the experimental method up to temperatures as high as 314° C (EHRLICH).

Table 3. *Typical Electrical Properties of Polytetrafluoroethylene*

Property	Units	Method	
Dielectric constant, 23° C, 60 to 10^9 cps		VON HIPPEL	2.0
Dielectric strength, short-time . . .	v/mil	D 149-44	400—500 [1]
Dissipation factor, 23° C, 60 to 10^9 cps		VON HIPPEL	$1 \pm 1 \times 10^{-4}$
Surface arc-resistance	sec	D 495-42	700 [2]
Volume resistivity, dry, 23° C . . .	ohm-cm	VON HIPPEL	10^{19}
Volume resistivity, 100% R.H. . .	ohm-cm	D 257-52T	10^{15}
Surface resistivity, 100% R.H.. . .	megohms	D 257-52T	3.6×10^6

Tests have been performed by A.S.T.M. methods unless otherwise indicated. Data shown are average values and should not be used for specifications.

[1] Value is 1000—2000 volts/mil in thickness of 5 to 12 mils.

[2] Does not track.

The dielectric constant over the same frequency range, however, decreases slightly, from 2.00 at 23° C to 1.81 at 314° C as expected from the thermal expansion (EHRLICH). The volume resistivity is 10^{19} ohm-cm at 23° C (VON HIPPEL) and is greater than 10^{15} ohm-cm even after prolonged soaking in water. The surface resistivity is 3.6×10^{12} ohms even at 100% relative humidity. Polytetrafluoroethylene also has great arc resistance. On exposure to an arc, regardless of time of exposure, the material leaves no carbonized conducting path since the only product of degradation is a nonconducting gas. Short-time dielectric strenghts are high and range from 500 volts/mil (1 mil = 10^{-3} in.) for sheets more than 100 mils thick to 4,000 volts/mil for very thin films. The dielectric strength is independent of temperature from 50 to 250° C.

In order to obtain the optimum dielectric strength, it is necessary to avoid the formation of voids. THOMAS, LONTZ, SPERATI, and MCPHERSON have shown that the dielectric strength may be as low as 200 volts/mil when polytetrafluoroethylene is fabricated in a way which produces several per cent voids. SUGITA, NAGAO, and TORIYAMA also showed how the elimination of voids improved the corona resistance of polytetrafluoroethylene and decreased the rate of dielectric breakdown. The dielectric breakdown under prolonged stress was studied by BRODHUN. He also showed how the presence of corona causes an additional effect. PAO and BJORKLUND performed quantum mechanical calculations of the energies of excited electronic states.

It has been mentioned in the section on viscoelasticity that dielectric loss maxima are sometimes observed at the same temperatures and frequencies as mechanical loss maxima. This correspondence has been particularly useful in the study of the Glass II transition (McCrum, 1959b). The electrical loss in these special regions of temperature and frequency is generally less than 10^{-3} and is detected only with the aid of special equipment. The source of the electrical loss in these transitions is not known but may involve the presence of a small number of extraneous dipoles. Some support for this suggestion comes from the fact that different workers do not always observe the same loss maxima. For instance, the loss at 160° C which was described by Krum and Muller was not observed in the work of Mikhailov, Kabin, and Smolyanskii.

VII. Chemical Properties

A. Resistance to Reagents and Solvents

Polytetrafluoroethylene is extremely resistant to attack by most highly corrosive chemicals. Materials such as aqua regia, hydrofluoric acid, fuming sulfuric acid, chlorosulfonic acid, hot fuming nitric acid, boiling solutions of sodium hydroxide, gaseous chlorine, hydrogen peroxide, and organic materials such as esters, ketones, alcohols, acid chlorides, and highly halogenated organic substances have no effect on the polymer (Renfrew and Lewis). While studying the attack of organic amines on hydrofluorocarbon polymers, Bro found that polytetrafluoroethylene was completely stable to these reagents. The only chemicals which are known to attack solid polytetrafluoroethylene are molten alkali metals, such as sodium or potassium, their solutions, chlorine trifluoride, and gaseous fluorine at elevated temperatures and pressures. There is no known solvent for the polymer at temperatures below about 300° C. Some highly fluorinated oils, however, will swell or dissolve polytetrafluoroethylene at temperatures near its crystalline melting point. (Doban, Sperati, and Sandt).

The almost universal chemical inertness of polytetrafluoroethylene has been attributed to the strength of the carbon-fluorine bond and the way in which the fluorine atoms protect the carbon chain from chemical attack (Doban, Sperati, and Sandt). From the theory of solubility, it is expected that the miscibility of hydrocarbons and fluorocarbons will be low. Experimental measurements indicate that the miscibility is even less than was expected from the theory. The possible explanations for this have been discussed by Scott.

Recently, the ability of thin films of polytetrafluoroethylene to transmit oxygen and carbon dioxide but not water has been used to construct a heart-lung machine (Crescenzi, Hofstra, Sze, Foster, and Claff).

B. Thermal Stability

Early study of the pyrolysis of polytetrafluoroethylene was reported by LEWIS and NAYLOR. A series of subsequent investigations have been described by SIEGLE, MUUS, and LIN and by workers at the National Bureau of Standards (FLORIN, WALL, BROWN, HYMO, and MICHAELSEN; MADORSKY, HART, STRAUSS, and SEDLAK; MICHAELSEN and WALL; WALL and MICHAELSEN).

The vacuum pyrolysis of thin samples of polytetrafluoroethylene follows first-order kinetics with monomer as the major decomposition product in the temperature range from 360—510° C (SIEGLE, MUUS, and LIN). The rate constant does not depend on either the molecular weight or the type of polymer and is characterized by an activation enthalpy of 83.0 kcal/mole and a frequency factor of 3×10^{19} sec^{-1}. The melt viscosity decreases during pyrolysis.

For thicker samples the rate of vacuum pyrolysis is controlled by diffusion of monomer. The course of the pyrolysis is changed by the presence of monomer or gaseous pyrolysis products. The rate of pyrolysis as measured by weight loss is reduced and there is a rapid decrease in melt viscosity at low conversion. According to SIEGLE, MUUS, and LIN, the most probable mechanism for the vacuum pyrolysis of polytetrafluoroethylene is random chain cleavage, propagation with a short kinetic chain length, and termination by disproportionation. The pressure of monomer in equilibrium with a chain radical increases from 4.8×10^{-3} mm Hg at 327° C to 7.6 mm at 510° C. One atmosphere of tetrafluoroethylene would be in equilibrium with a chain radical at 712°C.

C. Effects of Radiation

When polytetrafluoroethylene is exposed to ionizing radiation in the presence of oxygen, it degrades rather than crosslinks (WALL and FLORIN; CHARLESBY, 1954, 1955; RYAN). This is reflected in a decrease in melt viscosity by 10—100-fold (MATSUMAE, WATANABE, NISHIOKA, and ICHIMIYA) and an increase in electrical conductivity (FOWLER and FARMER). The decrease in apparent molecular weight is also suggested by the increase in density of irradiated polymer which has been heated above its melting point and recooled (NISHIOKA, MATSUMAE, WATANABE, TAJIMA, and OWAKI). WALL and FLORIN have shown that much of this degradation can be avoided if oxygen is excluded. It is believed that irradiation involves the breaking of carbon-fluorine bonds producing secondary radicals. In vacuum these radicals may combine to form crosslinks. They are, however, quite reactive toward other substances. Reaction with oxygen results in chain cleavage and further degradation with monomer as the primary product. Reaction with other olefins has been reported to give graft copolymers (CHEN, MESROBIAN, BALLANTINE, METZ, and GLINES; CHAPIRO; RESTAINO and REED).

VIII. Processing Technology

Above its melting point of 327° C, polytetrafluoroethylene has some properties more like a rubber than a liquid. The instantaneous Young's modulus is 2—3 × 10^7 dynes/cm^2, and the melt viscosity is about 10^{11} poises at 380° C (Nishioka and Watanabe). Because of this very high melt viscosity, it is not feasible to process the polymer by conventional extrusion or injection molding. Instead, techniques similar to those of powder metallurgy are employed. These involve three basic steps.

(1) The dry powder is compressed at moderate temperatures (e. g., 25—160° C) to produce a preform which has enough strength for careful handling but is easily broken.

(2) The preform is heated above the melting point, usually at about 380° C to allow the particles to coalesce. This is commonly called sintering.

(3) The polymer is cooled in a controlled manner to give the desired degree of crystallinity.

Thomas, Lontz, Sperati, and McPherson have discussed in detail the ways in which these three steps affect the quality of objects fabricated from polytetrafluoroethylene. It is important to use a relatively high pressure in making the preform to avoid the formation of voids, particularly in the case of granular polymer. These pressures vary from a few hundred to many thousand pounds per square inch, depending on the size of the particles. If the sintering temperature is too low and the time of sintering too short, there will not be adequate coalescence of the particles into a continuous mass. If, on the other hand, the sintering temperature is too high, the polymer will be degraded and the molecular weight will be reduced. The percent crystallinity is controlled by the rate at which the molding is cooled after sintering or by annealing slightly below the melting point. For maximum toughness, it is desirable to minimize crystallinity by cooling rapidly. Maximum stiffness is obtained by cooling very slowly.

Polytetrafluoroethylene powders can be extruded but not by the methods applied to conventional thermoplastics. The polymer must not be subjected to excessive mechanical working above the melting point or cracking of the finished piece will result. For this reason, the polymer is worked while it is still cold. The powder is compacted with the aid of a screw or a ram and forced through a hot die in which it is sintered (James). This method is used to fabricate rods or heavy wall tubing or to apply heavy coatings to electrical conductors.

The extrusion of lubricated powders, which has been described by Lontz, Jaffe, Robb, and Happoldt, is a unique method for fabricating polytetrafluoroethylene. An aqueous dispersion is coagulated by mechanical agitation. The powder is then dried and blended with 18—20% of a lubricant. Suitable lubricants include aliphatic and aromatic hydrocarbons, certain alcohols, esters, and silicones. The resulting paste is passed through an extruder at a temperature below the melting point to form a tape, wire coating, or other shape. The lubricant is removed by heating below the melting point, and the product is then sintered and either annealed or quenched to give the desired level of crystallinity.

The rheology of lubricated polytetrafluoroethylene compositions was studied by LEWIS and WINCHESTER. The mechanism appeared to be a combination of permanent and elastic deformations in the region just before the orifice of the die in the extruder. As a result of permanent deformation, the polymer particles are partially transformed into long fibers. The relative amounts of permanent and recoverable deformation were related to the rate and temperature of extrusion and the geometry of the extruder. Plastic deformation is favored by extruding at temperatures above the 19 and 30° transitions (SNELLING and LONTZ).

Films of polytetrafluoroethylene can also be made by casting an aqueous dispersion on a supporting surface, drying to remove the water, baking, and cooling the baked film. (LONTZ and HAPPOLDT). If the film is too thick, cracks are formed which cannot be eliminated by fusing, even on prolonged baking. This critical thickness is approximately 40 μ and limits the amount that can be laid down in a single application. Thicker films are formed by casting and baking a series of layers on top of each other. It is necessary to bake the film after casting each layer. In a similar way, it is possible to impregnate woven fabrics and mat structures to render them chemically more resistant. Aqueous dispersions of polytetrafluoroethylene are also used to form supported films on substrates such as metal, glass, and ceramics (ELLIOTT). Such coatings are useful because of their chemical resistance, electrical properties, and lubricity (FITZSIMMONS and ZISMAN).

The methods used for processing polytetrafluoroethylene while different from those generally used with polymeric materials, have made the polymer readily accessible in large and complex shapes at reasonable cost. The principles underlying these methods are now available for use in studying other seemingly intractable polymers.

Bibliography

ALLAN, A. J. G.: Wettability and friction of polytetrafluoroethylene film: effect of prebonding treatments. J. Polymer Sci. **24**, 461–466 (1957).

— Plastics as solid lubricants and bearings. Lubrication Eng. **14**, 211–215 (1958).

— and F. M. CHAPMAN: Frictional properties of TFE-fluorocarbon resins. Materials in Design Eng. **48**, 106–108 (1958).

— and R. ROBERTS: Wettability of perfluorocarbon polymer films: effect of roughness. J. Polymer Sci. **39**, 1–8 (1959).

BEECROFT, R. I., and C. A. SWENSON: Behavior of polytetrafluoroethylene (Teflon) under high pressures. J. Appl. Phys. **30**, 1793—1798 (1959).

BERNETT, M. K., and W. A. ZISMAN: Wetting of low-energy solids by aqueous solutions of highly fluorinated acids and salts. J. Phys. Chem. **63**, 1911—1916 (1959).

BERRY, K. L., and J. H. PETERSON: Tracer studies of oxidation-reduction polymerization and molecular weight of Teflon tetrafluoroethylene resin. J. Am. Chem. Soc. **73**, 5195—5197 (1951).

BOWDEN, F. P.: Frictional properties of porous metal impregnated with plastic. Research (London) **3**, 147—168 (1950).

— Friction on snow and ice. Proc. Roy. Soc. **A 217**, 462—478 (1953).

— Friction on snow and ice and the development of some fast-running skiis. Nature (London) **176**, 946—947 (1955).

BRANDT, W.: Calculation of compressibilities of high polymers from the energy of interaction between chains. J. Chem. Phys. **26**, 262—270 (1957).

BRIDGMAN, P. W.: Rough compressions of 177 substances to 40,000 kg/cm². Proc. Am. Acad. Arts Sci. **76**, 71—74 (1948).

BRO, M. I.: The stability of fluorine-containing polymers to amines. J. Appl. Polymer Sci. **1**, 310—312 (1959).

— and C. A. SPERATI: End groups in tetrafluoroethylene polymers. J. Polymer Sci. **38**, 289—205 (1959).

BRODHUN, C. G.: Dielectric breakdown of polymers under prolonged stress. Phys. Rev. **86**, 653 (1952).

BRUNAUER, S., P. H. EMMETT and E. TELLER: Adsorption of gases in multimolecular layers. J. Am. Chem. Soc. **60**, 309—319 (1938).

BRYANT, W. M. D.: Free energies of formation of fluorocarbons and their radicals. Thermodynamics of formation and depolymerization of polytetrafluoroethylene. J. Polymer Sci., in press.

BUNN, C. W., and E. R. HOWELLS: Structures of molecules and crystals of fluorocarbons. Nature (London) **174**, 549—551 (1954).

— A. J. COBBOLD and R. P. PALMER: The fine structure of polytetrafluoroethylene. J. Polymer Sci. **28**, 365—376 (1958).

CHAPIRO, A. : Preparation of graft copolymers of polytetrafluoroethylene by radiochemistry. J. Polymer Sci. **34**, 481—501 (1959).

CHARLESBY, A.: How radiation affects long-chain polymers. Nucleonics **12**, 18—25 (1954).

— Some radiation effects in long chain polymers. Plastics Institute Trans. **23**, 133—138 (1955).

CHEN, W. K. W., R. B. MESROBIAN, D. S. BALLANTINE, D. J. METZ and A. GLINES: Studies of graft copolymers derived by ionizing radiation. J. Polymer Sci. **23**, 903—913 (1957).

CLARK, E. S., and L. T. MUUS: Unusual features in the crystal structure of polytetrafluoroethylene. Meeting of the American Chemical Society, New York, September 1957.

— — Intensity of bragg reflections and disorder in crystalline polymers. Meeting of the American Crystallographic Association, Milwaukee, Wisconsin, June 23, 1958.

— The helical structure concept and synthetic polymers. Symposium on "Helices in macromolecular systems" at Polytechnic Institute of Brooklyn, May 16, 1959.

CRESCENZI, A. A., P. C. HOFSTRA, K. C. SZE, B. H. FOSTER and C. L. CLAFF: Development of a simplified disposable membrane oxygenator. 45th Annual Clinical Congress of the American College of Surgeons, Atlantic City, September 28, 1959.

DOBAN, R. C., C. A. SPERATI and B. W. SANDT: The physical properties of "Teflon" polytetrafluoroethylene. Soc. Plastics Engrs Journal **11**, 17–21, 24, 30 (1955).

– A. C. KNIGHT, J. H. PETERSON and C. A. SPERATI: The molecular weight of polytetrafluoroethylene. Meeting of the American Chemical Society, Atlantic City, September 1956.

– Method for the preparation of cementable fluorocarbon polymer surfaces. U. S. Patent 2,871,144 (Jan. 27, 1959) assigned to E. I. du Pont de Nemours & Co.

DOWNING, F. B., A. F. BENNING and R. C. MCHARNESS: Octafluorocyclobutane and pyrolytic process for its production. U. S. Patent 2,384,821 (Sept. 18, 1945) U. S. Patent 2,387,247 (Oct. 23, 1945).

DUUS, H. C.: Thermochemical studies on fluorocarbons. Ind. Eng. Chem. **47**, 1445–1449 (1955).

DYMENT, J., and H. ZIEBLAND: Tensile properties of plastics at low temperatures. J. Appl. Chem. (London) 8, 203–206 (1958).

EHRLICH, P.: Dielectric properties of "Teflon" from room temperature to 314° C and from frequencies of 10^2 to 10^5 cps. J. Research Natl. Bur. Standards **51**, 185–188 (1953).

ELLIOTT, E. M.: Polytetrafluoroethylene dispersions for metal finishing. Trans. Inst. Metal Finishing **33**, 355–365 (1956).

FITZSIMMONS, V. G., and W. A. ZISMAN: Thin films of polytetrafluoroethylene resin ("Teflon") as lubricants and preservative coatings for metals. NRL Report 4753. Naval Research Laboratory, Washington, D. C., June 15, 1956. See also Ind. Eng. Chem. **50**, 781–784 (1958).

FLORIN, R. E., L. A. WALL, D. W. BROWN, L. A. HYMO and J. D. MICHAELSEN: Factors affecting the thermal stability of polytetrafluoroethylene. J. Research Natl. Bur. Standards **53**, 121–130 (1954).

FOWLER, J. F., and F. T. FARMER: Conductivity induced in polytetrafluoroethylene by x-rays. Nature (London) **174**, 136–137 (1954).

FOX, H. W., and W. A. ZISMAN: The spreading of liquids on low energy surfaces. I. polytetrafluoroethylene. J. Colloid Sci. **5**, 514–531 (1950).

FURUKAWA, G. T., R. E. MCCOSKEY and G. J. KING: Calorimetric properties of polytetrafluoroethylene ("Teflon") from 0° to 365° K. J. Research Natl. Bur. Standards **48**, 273–278 (1952).

HALS, L. T., T. S. REID, and G. H. SMITH: The preparation of terminally unsaturated perfluoroolefins by the decomposition of the salts of perfluoro acids. J. Am. Chem. Soc. **73**, 4054 (1951).

HANFORD, W. E., and R. M. JOYCE: Polytetrafluoroethylene. J. Am. Chem. Soc. **68**, 2082–2085 (1946).

HASZELDINE, R. N.: The addition of free radicals to unsaturated systems. IV. the direction of radical addition to hexafluoropropene. J. Chem. Soc. 3559–3564 (1953).

HIPPEL, A. R. VON: Dielectric materials and applications. New York: Technology Press of MIT and John Wiley 1954.

HYNDMAN, D., and G. F. ORIGLIO: NMR absorption in "Teflon" fibers. J. Applied Physics **31**, 1849—1852 (1960).

ILLERS, K. H., and E. JENCKEL: Mechanische Relaxationserscheinungen in Polytetrafluoräthylen. Kolloid-Z. **160**, 97–106 (1958).

JAMES, D. D.: Molding and extrusion of "Teflon". India Rubber World **124**, 68–70 (1951).

KABIN, S. P.: On the dynamic mechanical properties of polyethylene and polytetrafluoroethylene. Soviet Physics Technical Physics **1**, 2542–2546 (1956).

KILIAN, H. G., and E. JENCKEL: The melting and crystallization of some high-polymeric substances as observed by x-ray analysis. Z. Elektrochem. **63**, 308—321 (1959).

KIYAMA, R., J. OSUGI and S. KUSUHARA: Studies on explosive reactions of tetrafluoroethylene and acetylene with oxygen or air. Rev. Phys. Chem. Jap. **27**, 22—41 (1957).

KRUM, F., and F. H. MULLER: Vorbehandlung und dielektrisches Verhalten Hochpolymerer. Kolloid-Z. **164**, 81—107 (1959).

LEKSINA, I. E., and S. I. NOVIKOVA: Thermal expansion of fluoroplast between —190 and 325° C. Sov. Phys. Solid State **1**, 453—459 (1959).

LEWIS, E. E., and M. A. NAYLOR: Pyrolysis of polytetrafluoroethylene. J. Am. Chem. Soc. **69**, 1968—1970 (1947).

— and C. M. WINCHESTER: Rheology of lubricated polytetrafluoroethylene compositions — equipment and operating variables. Ind. and Eng. Chem. **45**, 1123—1127 (1953).

LIANG, C. Y., and S. KRIMM: Infrared spectra of high polymers III. polytetrafluoroethylene and polychlorotrifluoroethylene. J. Chem. Physics **25**, 563—571 (1956).

LOCKE, E. G., W. R. BRODE and A. L. HENNE: Fluorochloroethanes and fluorochloroethylenes. J. Am. Chem. Soc. **56**, 1726—1728 (1934).

LONTZ, J. F., and W. B. HAPPOLDT JR.: "Teflon" tetrafluoroethylene resin dispersion. A new aqueous colloidal dispersion of polytetrafluoroethylene. Ind. and Eng. Chem. **44**, 1800—1805 (1952).

— J. A. JAFFE, L. E. ROBB and W. B. HAPPOLDT JR.: "Teflon" tetrafluoroethylene resin dispersion-extrusion properties of lubricated resin from coagulated dispersion. Ind. and Eng. Chem. **44**, 1805—1810 (1952).

LUPTON, J. M.: Effect of pressure on the specific volume of polymer melts. Meeting of the American Chemical Society, Chicago, September 1958.

MADORSKY, S. L., V. E. HART, S. STRAUSS and V. A. SEDLAK: Thermal degradation of tetrafluoroethylene and hydrofluoroethylene polymers. J. Research Natl. Bur. Standards **51**, 327—333 (1953).

MARX, P., and M. DOLE: Specific heat of synthetic high polymers. V. A study of the order-disorder transition in polytetrafluoroethylene. J. Am. Chem. Soc. **77**, 4771—4774 (1955).

MATSUMAE, K., M. WATANABE, A. NISHIOKA and T. ICHIMIYA: Viscosity and elasticity of gamma-irradiated polytetrafluoroethylene resin above the melting point. J. Polymer Sci. **28**, 653—655 (1958).

MCCRUM, N. G.: An internal friction study of polytetrafluoroethylene. J. Polymer Sci. **34**, 355—369 (1959a).

— A study of internal friction in copolymers of tetrafluoroethylene and hexafluoropropylene. Die Makromol. Chem. **34**, 50—66 (1959b).

— Torsion pendulum method for determining crystallinity and void content of tetrafluoroethylene resins. ASTM Bull. No. **242**, 80—82 (1959c).

MCGEER, P. L., and H. C. DUUS: Effect of pressure on the melting point of "Teflon" tetrafluoroethylene resin. J. Chem. Phys. **20**, 1813—1814 (1952).

MCGREW, F. C.: Some structural investigations on polytetrafluoroethylene. International Symposium on Fluorine Chemistry, Birmingham, England, July 17, 1959.

MICHAELSEN, J. D., and L. A. WALL: Further studies on the pyrolysis of polytetrafluoroethylene in the presence of various gases. J. Research Natl. Bur. Standards **58**, 327—331 (1957).

MIKHAILOV, G. P., S. P. KABIN and A. L. SMOLYANSKII: Dielectric losses of polytetrafluoroethylene. J. Techn. Phys. (U.S.S.R.) **25**, 2179—2182 (1955).

MIYAKE, A.: On the rotational transition of linear high polymer crystals. Chem. High Polymers (Tokyo) **15**, 153—159 (1958).

MOYNIHAN, R. E.: The molecular structure of perfluorocarbon polymer: IR studies on polytetrafluoroethylene. J. Am. Chem. Soc. **81**, 1045—1050 (1959).

NAGAMATSU, K., T. YOSHITOMI and T. TAKEMOTO: On the viscoelastic properties of crystalline polymers I. polytetrafluoroethylene. J. Colloid Sci. **13**, 257—265 (1958).

NELSON, E. R., T. J. KILDUFF and A. A. BENDERLY: Bonding of "Teflon". Ind. Eng. Chem. **50**, 329—330 (1958).

NEUGEBAUER, C. A., and J. L. MARGRAVE: The heats of formation of tetrafluoromethane and 1,1-difluoroethylene. J. Phys. Chem. **60**, 1318—1321 (1956).

NISHIOKA, A., K. MATSUMAE, M. WATANABE, M. TAJIMA and M. OWAKI: Effects of gamma radiation on some physical properties of polytetrafluoroethylene resin. J. Appl. Polymer Sci. **2**, 114—119 (1959).

— M. TAJIMA, and M. OWAKI: Melt fracture and crystallization of gamma irradiated polytetrafluoroethylene. J. Polymer Sci. **28**, 617—619 (1958).

— and M. WATANABE: Viscosity and elasticity of polytetrafluoroethylene resin above the melting point. J. Polymer Sci. **24**, 298—300 (1957).

PAO, Y. H., and R. F. BJORKLUND: Wave functions and energy eigenvalues for the polytetrafluoroethylene chain. J. Applied Physics **31**, 1925—1934 (1960).

PIERCE, R. H. H., JR., E. S. CLARK, J. F. WHITNEY and W. M. D. BRYANT: Crystal structure of polytetrafluoroethylene. Meeting of the American Chemical Society Atlantic City, September, 1956.

PLUNKETT, R. J.: Tetrafluoroethylene polymers. U. S. Patent 2,230,654 (Feb. 4, 1941).

POWLES, J. G., and J. A. E. KAIL: Nuclear magnetic resonance absorption in various polytetrafluoroethylenes. J. Polymer Sci. **31**, 183—187 (1958).

PURVIS, R. J., and W. R., and W. R. BECK: Method of activating the surface of perfluorocarbon polymers and resultant article .U. S. Patent 2,789,063 (April 16, 1957) — assigned to Minnesota Mining and Mfg. Co.

QUINN, F. A., JR., D. E. ROBERTS and R. N. WORK: Volume-temperature relationships for the room temperature transition in "Teflon". J. Appl. Physics **22**, 1985—1086 (1951).

RAPPAPORT, G.: Method of bonding a fluorinated synthetic resin to another material. U. S. Patent 2,809,130 (Oct. 8, 1957) — assigned to General Motors Corp.

RENFREW, M. M., and E. E. LEWIS: Polytetrafluoroethylene: heat-resistent, chemically inert plastic. Ind. Eng. Chem. **38**, 870—877 (1946).

— Polymerization of tetrafluoroethylene with dibasic acid peroxide catalysts. U. S. Patent 2,534,058 (Dec. 12, 1950).

RESTAINO, A. J., and W. N. REED: Kinetics of the gamma-induced graft copolymerization of vinyl acetate to "Teflon". J. Polymer Sci. **36**, 499—510 (1959).

RIGBY, H. A., and C. W. BUNN: A room-temperature transition in polytetrafluoroethylene. Nature (London) **164**, 583 (1949).

RILEY, M. W.: Selection and design of fluorocarbon plastics. Materials and Methods Manual No. 138, 129—148 (June 1957).

ROEDEL, M. J.: The molecular structure of polyethylene, I. chain branching in polyethylene during polymerization. J. Am. Chem. Soc. **75**, 6110—6113 (1953).

RUFF, O., and O. BRETSCHNEIDER: Die Bildung von Hexafluoräthan und Tetrafluoräthylen aus Tetrafluorkohlenstoff. Z. anorg. Chem. **210**, 173—183 (1933).

RYAN, J. W.: Radiation of polytetrafluorethylene. Modern Plastics **31**, 152 (Oct. 1953).

RYLAND, A. L.: X-ray diffraction. J. Chem. Ed. **35**, 80–83 (1958).

SCOTT, R. L.: The anomalous behavior of fluorocarbon solutions. J. Phys. Chem. **62**, 136–145 (1958).

SHOOTER, K. V., and D. TABOR: The frictional properties of plastics. Proc. Phys. Soc. B **65**, 661–671 (1952).

— and P. H. THOMAS: Frictional properties of some plastics. Research (London) **2**, 533–535 (1949).

SIEGLE, J. C., L. T. MUUS and T. P. LIN: Pyrolysis of polytetrafluoroethylene. Meeting of the American Chemical Society, Atlantic City, September 1956.

SINNOTT, K. M.: Paper in preparation.

SLICHTER, W. P.: Nuclear magnetic resonance in some fluorine derivatives of polythene. J. Polymer Sci. **24**, 173–188 (1957).

SMITH, J. A. S.: A nuclear resonance investigation of polytetrafluoroethylene. Discussions Faraday Soc. No. **19**, 207–215 (1955).

SNELLING, G. R., and J. F. LONTZ: Mechanism of lubricant extrusion of "Teflon" TFE-tetrafluoroethylene resins. J. Appl. Polymer Sci. **3**, 257–265 (1960).

SPERATI, C. A., and J. L. MCPHERSON: The effect of crystallinity and molecular weight on physical properties of polytetrafluoroethylene. Meeting of the American Chemical Society, Atlantiv City, September 1956.

STARKWEATHER, H. W., JR., and R. H. BOYD: The entropy of melting of some linear polymers. J. Phys. Chem. **64**, 410–414 (1960).

— Heat capacity of chain polymers at low temperatures. J. Polymer Sci. **45**, 525—527 (1960).

SUGITA, K., S. NAGAO and Y. TORIYAMA: The corona resisting property of polytetrafluoroethylene. Brit. J. Appl. Phys. **7**, 38 (1956).

SWENSON, A. A.: Mechanical properties of "Teflon" at low temperatures. Rev. Sci. Instr. **25**, 834 (1954).

TERANISHI, H.: Studies on the explosions under high pressures, IV. The explosions of tetrafluoroethylene mixed with oxygen or air. Rev. Phys. Chem. Japan **28**, 9–23 (1958).

THOMAS, P. E., J. F. LONTZ, C. A. SPERATI and J. L. MCPHERSON: Effects of fabrication on the properties of "Teflon" resins. Soc. Plastics Engrs. **12**, (June 1956).

TORKINGTON, P., and H. W. THOMPSON: The infrared spectra of fluorination hydrocarbons I. Trans. Faraday Soc. **41**, 236–246 (1945).

WALL, L. A., and J. D. MICHAELSEN: Thermal decomposition of polytetrafluoroethylene in various gaseous atmospheres. J. Research Natl. Bur. Standards **56**, 27–34 (1956).

— and R. E. FLORIN: Polytetrafluoroethylene — a radiation resistant polymer. J. Appl. Polymer Sci., **2**, 251 (1959).

WEIR, C. E.: Compressibility of natural and synthetic high polymers at high pressures. J. Research Natl. Bur. Standards **46**, 207–212 (1051).

— Transitions and phases of polytetrafluoroethylene ("Teflon"). J. Research Natl. Bur. Standards **50**, 95–97 (1953).

WILSON, C. W., and G. E. PAKE: Nuclear magnetic relaxation in polytetrafluoroethylene and polythene. J. Chem. Phys. **27**, 115–122 (1957).

Fortschr. Hochpolym.-Forsch., Bd. 2, S. 496—577 (1961)

Kettenübertragung bei der radikalischen Polymerisation

Von

G. Henrici-Olivé und S. Olivé

Monsanto Research S. A. Zürich, Schweiz

Mit 26 Abbildungen

Inhaltsverzeichnis

Einleitung

Seit FLORY (*44*) 1937 das Prinzip der Kettenübertragung in die Polymerisationskinetik einführte, und BREITENBACH (*27*) (1938) und G. V. SCHULZ (*126*) (1939) zuerst über die Verminderung des Polymerisationsgrades bei Gegenwart von Lösungsmitteln berichteten, ist von einer großen Anzahl von Autoren die kettenübertragende Wirkung der verschiedensten Substanzen untersucht worden.

Diese Untersuchungen haben einerseits das praktische Ziel, geeignete Substanzen zu finden, mit deren Hilfe man das Molekulargewicht eines Polymerisates beliebig regeln kann. Andererseits sind sie aber auch von grundlegendem wissenschaftlichem Interesse: Durch Vergleich der kettenübertragenden Wirkung verschiedener Substanzen gegenüber einer Reihe von Polymeren kann man Schlüsse ziehen sowohl auf die Reaktivität der Polymerradikale, als auch auf die Bindungsfestigkeit der übertragenen Atome (oder Atomgruppen) in der Überträgersubstanz.

Leider sind bei sehr vielen der publizierten Arbeiten einige wesentliche Punkte nicht richtig, oder überhaupt nicht, berücksichtigt worden, wodurch die Vergleichbarkeit der von verschiedenen Autoren bestimmten Übertragungskonstanten stark beeinträchtigt wird.

Es handelt sich hierbei zunächst um die Tatsache, daß die Bruttogeschwindigkeit sehr häufig bei Gegenwart eines Kettenüberträgers vermindert wird. Dies kann durch eine Verringerung der Startgeschwindigkeit (Radikalausbeute des Initiators) oder aber auch durch eine zusätzliche Abbruchsreaktion verursacht sein. In jedem Fall muß die Verminderung der Geschwindigkeit jedoch berücksichtigt werden.

Ferner spielt die Mittelwertbildung der experimentellen Molekulargewichtsbestimmung eine wichtige Rolle. Die kinetischen Beziehungen, mit deren Hilfe Übertragungskonstanten bestimmt werden, sind in ihrer ursprünglichen Form immer für den Zahlendurchschnitt des Polymerisationsgrades abgeleitet. Die üblicherweise verwendeten Methoden zur Molekulargewichtsbestimmung liefern jedoch das Gewichtsmittel (Lichtzerstreuung) oder das von letzterem nicht sehr verschiedene Viscositätsmittel (Viscosität, Geschwindigkeitsultrazentrifuge). Bei der Umrechnung dieser Durchschnittswerte auf das Zahlenmittel muß die Molekulargewichtsverteilung im Polymerisat in Betracht gezogen werden. In allen Polymeren, in welchen der Kettenabbruch ganz oder teilweise durch Kombination erfolgt, ist die Molekulargewichtsverteilung jedoch von dem Ausmaß der Kettenübertragung abhängig, d. h. von Fall zu Fall verschieden. Es ist daher nicht möglich, die Uneinheitlichkeit des Polymeren durch eine einmalige Eichung der Viscosität etwa durch die das Zahlenmittel liefernde Osmose zu berücksichtigen. Wenn man bedenkt, daß für vollständig durch Kombination abgebrochene Moleküle

$\overline{P}_w/\overline{P}_n = 1{,}5$, hingegen für stark übertragene Polymere $\overline{P}_w/\overline{P}_n = 2$, so erkennt man, daß in ungünstigen Fällen auf diese Weise Fehler bis zu 30% im Polymerisationsgrad entstehen können (vgl. Abschnitt II).

Es erschien daher wünschenswert, dem heutigen Stand der Kenntnis entsprechend die formalkinetischen Grundlagen zusammenfassend darzustellen, mit deren Hilfe man Übertragungskonstanten bestimmen kann, unter Berücksichtigung der veränderlichen Molekulargewichtsverteilung und etwaiger Änderungen in der Geschwindigkeit. In wesentlichen Grundzügen stützen wir uns dabei auf eine bereits veröffentlichte Arbeit (*130*).

Zur Erleichterung der Anwendung der abgeleiteten Gleichungen soll eine tabellarische Zusammenfassung dienen, welche am Ende des III. Abschnittes angefügt wurde, und mit deren Hilfe für jeden Spezialfall die zur Auswertung erforderliche Gleichung zusammengestellt werden kann.

Anschließend wird die Kettenübertragung an den drei Beispielen Styrol (Kombinationsabbruch), Methylmethacrylat (Kombination und Disproportionierung) und Vinylacetat (Kombinationsabbruch vernachlässigbar) behandelt. Hierzu werden großenteils Literaturdaten verwendet, wobei, sofern dies notwendig erschien, die Übertragungskonstanten mit Hilfe der abgeleiteten Gleichungen neu berechnet wurden. Daß hierbei kein Anspruch auf Vollständigkeit erhoben werden kann, ergibt sich aus der Vielzahl der Publikationen auf diesem Gebiet.

I. Ausgangsposition

a) Die „Mayo-Gleichung" und ihre beiden Voraussetzungen

Bei den meisten Übertragungsreaktionen handelt es sich um die Übertragung eines Wasserstoffatoms von der Überträgersubstanz auf die wachsende Polymerkette, wodurch die Radikalstelle an letzterer abgesättigt, an der Überträgersubstanz jedoch ein neues Radikal gebildet wird:

$$P^* + HX \rightarrow PH + X^* \tag{I, 1}$$

Es sind allerdings auch Übertragungen anderer Atome (z. B. Halogene) bekannt.

Schulz und Blaschke (*123*) haben 1942 zuerst die kinetische Behandlung eines Übertragungseffektes gegeben. An dem Beispiel des Methylmethacrylates haben sie demonstriert, wie aus der Verminderung des Polymerisationsgrades die Übertragungskonstante des *Monomeren* bestimmt werden kann.

Wenig später (1943) haben Gregg und Mayo (*49*) eine systematische kinetische Behandlung des Übertragungseffektes von *Lösungsmitteln* gegeben. Nach der von Mayo eingeführten Methode trägt man den Reziprokwert des Polymerisationsgrades graphisch auf als Funktion des

Verhältnisses der Konzentrationen von Monomerem und Lösungsmittel und erhält die Übertragungskonstante als Steigung einer geraden Linie nach der Gleichung:

$$\frac{1}{\overline{P}_n} = \frac{1}{\overline{P}_{n,0}} + C\,\frac{[LM]}{[M]}\,. \qquad \text{(I,2)}$$

Hierbei sind $[LM]$ und $[M]$ die Konzentrationen des übertragenden Lösungsmittels und des Monomeren in Mol/l, C ist das Verhältnis der Geschwindigkeitskonstanten von Kettenübertragung und Kettenwachstum ($C \equiv k_{ü}/k_w$), und $\overline{P}_{n,0}$ und $\overline{P}_n$ sind die Polymerisationsgrade bei Abwesenheit bzw. Gegenwart des betreffenden Lösungsmittels.

Gleichung I,2 ist in der Folge immer wieder — teilweise nach verschiedenen Umformungen — zur Bestimmung von Übertragungskonstanten herangezogen worden. Dabei ist jedoch häufig übersehen worden, daß diese Gleichung zwei Voraussetzungen hat.

Die erste Voraussetzung ist, daß der erste Term rechts tatsächlich ein konstanter Wert sein muß. Daß dies keineswegs immer der Fall ist, wird am besten klar, wenn man auf den Ursprung von Gl. I,2 zurückgreift.

Der Polymerisationsgrad *ohne* Übertragung ist bekanntlich gegeben durch das Verhältnis der Wahrscheinlichkeiten von Kettenwachstum und Abbruch. Da die Reaktionsgeschwindigkeiten proportional den Wahrscheinlichkeiten sind, ist also

$$\overline{P}_{n,0} = k\,\frac{v_w}{v_{ab}}\,. \qquad \text{(I,3)}$$

v_w und v_{ab} bedeuten hierbei die Geschwindigkeiten von Wachstum und Abbruch; der Faktor k ist notwendig, falls der gegenseitige Abbruch zweier wachsender Ketten durch Kombination erfolgt. In diesem Falle werden zwei Einzelketten miteinander zu einem Molekül verbunden; es ist dann $k = 2$. Für Disproportionierungsabbruch ist $k = 1$.

Sofern Übertragung vorliegt, werden auch durch diese Reaktion Moleküle abgebrochen (siehe Gl. I,1). Es ist daher

$$\overline{P}_n = \frac{v_w}{(v_{ab}/k) + v_{ü}}\,. \qquad \text{(I,4)}$$

In die Gleichungen I,3 und I,4 kann man die entsprechenden Ausdrücke für die Geschwindigkeiten einsetzen (vgl. I,11, 12 und 15). Man erhält dann, wenn man außerdem zur Trennung der einzelnen Terme die Reziprokwerte verwendet:

$$\frac{1}{\overline{P}_{n,0}} = \frac{1}{k}\cdot\frac{k_{ab}\,v_{Br}}{k_w^2\,[M]^2} \qquad \text{(I,5)}$$

und

$$\frac{1}{\overline{P}_n} = \frac{1}{k}\cdot\frac{k_{ab}\,v_{Br}}{k_w^2\,[M]^2} + \frac{k_{ü}\,[X]}{k_w\,[M]}\,. \qquad \text{(I,6)}$$

Ein Vergleich dieser beiden Gleichungen mit (I,2) zeigt, daß (I,2) und (I,6) identisch sind, sofern man in letzterer anstelle des beliebigen Überträgers X das Lösungsmittel einsetzt.

Der erste Term rechts in Gl. I,6 ist nun nur in gewissen Fällen konstant. Zum Beispiel ist bei der thermischen Polymerisation in einer Reihe von Lösungsmitteln die Bruttogeschwindigkeit nahezu proportional dem Quadrat der Monomerkonzentration (*49, 51*). Dann kann Gl. I,2 in der dort angegebenen Form ausgewertet werden. Sobald jedoch Verzögerung eintritt ($v_{Br} \sim [M]^n; n > 2$), sind die so erhaltenen Übertragungskonstanten nicht mehr zuverlässig (siehe Abschnitt III, e).

Bei Anwendung eines radikalliefernden Initiators (I) ist erfahrungsgemäß

$$v_{Br} \sim [M]^n [I]^{1/2} \tag{I,7}$$

mit $1 \gtreqless n < 3/2$.

Es ist daher nicht möglich, das erste Glied rechts in Gl. I,2 bzw. I,6 dadurch konstant zu halten, daß man bei konstantem Verhältnis $[I]/[M]$ oder bei konstantem Verhältnis $[I]^{1/2}/[M]$ arbeitet, wie dies von verschiedenen Autoren versucht wird [vgl. z. B. KAPUR (*67*), MISRA u. Mitarb. (*34, 93, 36*), SANTHAPPA u. Mitarb. (*112*), THOMPSON u. Mitarb. (*143*)]. In beiden Fällen wird die Radikalausbeute des Initiators, welche die Ursache für den gebrochenen Exponenten in Gl. I,7 ist, nicht richtig berücksichtigt.

PALIT (*101*) hat diese Schwierigkeit umgangen, indem er den ersten Term rechts in Gl. I,6 aus den Versuchsdaten ausrechnet und mit seinem Zahlenwert einsetzt. Dies ist sicher das aussichtsreichste Verfahren und wird im wesentlichen auch in der vorliegenden Arbeit verwendet. Es versagt lediglich dann, wenn der Kettenüberträger gleichzeitig ein Polymerisationsverzögerer ist, d. h. wenn die nach Gl. I,1 entstehenden Überträgerradikale einen zusätzlichen Kettenabbruch verursachen (siehe Abschnitt III, e).

Die zweite Voraussetzung für die Anwendbarkeit von Gl. I,2 bzw. I,6 ist, daß die verwendeten Polymerisationsgrade *Zahlenmittel* ($\overline{P}_n$) darstellen müssen. Nur in den seltensten Fällen steht jedoch wirklich der Zahlendurchschnitt zur Verfügung.

Die am häufigsten verwendete Meßgröße zur Molekulargewichtsbestimmung ist die Viscositätszahl $[\eta]$, da sie sich mit verhältnismäßig geringem Aufwand an Zeit und Material mit großer Genauigkeit bestimmen läßt. Die Viscositätsbestimmung ist aber bekanntlich keine Absolutmethode, sondern muß mit einer anderen Methode der Molekulargewichtsbestimmung geeicht werden.

Bei einer polymolekularen Substanz ist zwischen zwei Größen jedoch nur dann ein eindeutiger Zusammenhang möglich, wenn sie demselben Durchschnittswert entsprechen. MEYERHOFF (*89*) hat darauf hingewiesen, daß aus diesem Grunde das osmotische Molekulargewicht mit seinem Zahlendurchschnitt zur Eichung der Viscosität weniger geeignet ist, da letztere einen Mittelwert liefert, welcher nicht sehr verschieden vom Gewichtsmittel ist. Diese Tatsache muß besonders dann berücksichtigt werden, wenn die untersuchten Präparate verschiedene Molekulargewichtsverteilungen aufweisen. Bei Polymeren, deren normaler Kettenabbruch durch Kombination zweier wachsender Ketten erfolgt, führt z. B. jede Kettenübertragung notwendigerweise zu einer Veränderung der Verteilung.

Eindeutig definierte Mittelwerte des Polymerisationsgrades lassen sich mit Hilfe der Viscositätszahl grundsätzlich nur in folgenden Fällen erhalten:

1. Die Eichung erfolgt mit idealen Fraktionen, für welche $P_n = P_\eta = P_w$. In diesem Fall ist es gleichgültig, mit welcher Methode geeicht wird. Die Anwendung der Eichkurve bzw. Eichgleichung auf unfraktionierte Polymere liefert dann immer den Durchschnittswert, den die Viscosität liefert und welchen wir als *Viscositätsmittel* ($\overline{P}_\eta$) bezeichnen. Dieser Fall ist jedoch praktisch nur angenähert realisierbar, da die üblichen Fraktioniermethoden keine idealen Fraktionen liefern (siehe Abschnitt IV, c).

2. Man eicht mit einer Methode, die denselben Mittelwert liefert, wie die Viscosität. Dies ist nach MEYERHOFF (*88*) mit guter Näherung der Fall für die Molekulargewichtsbestimmung mit der Geschwindigkeits-Ultrazentrifuge in Verbindung mit der Diffusion. Die unvermeidliche Uneinheitlichkeit der zur Eichung verwendeten Fraktionen spielt dann keine Rolle. Die Anwendung der Eichgleichung auf die Viscositätszahl von unfraktionierten Polymeren liefert wiederum das *Viscositätsmittel* ($\overline{P}_\eta$). Vom experimentellen Standpunkt ist das auf diese Weise bestimmte Molekulargewicht eines Polymeren heute das zuverlässigste, da sowohl die Messung der Viscositätszahl, als auch jene von Sedimentation und Diffusion mit großer Genauigkeit durchgeführt werden können (siehe Abschnitt IV, c).

3. Die Eichung erfolgt mit unfraktionierten Polymeren; Eichpräparate und später zu messende Präparate haben jedoch die gleiche Molekulargewichtsverteilung. Man erhält dann den Mittelwert der Eichmethode. Auf diese Weise kann z. B. bei Polymeren, die durch Disproportionierung abbrechen, das Zahlenmittel erhalten werden, sofern man mit Osmose eicht. Man hat jedoch die Unsicherheit osmotischer Molekulargewichtsbestimmungen an unfraktionierten Polymeren (*120, 91, 132*) in Kauf zu nehmen.

Der dritte, neben Zahlen- und Viscositätsmittel gebräuchliche Durchschnittswert des Polymerisationsgrades ist das *Gewichtsmittel* ($\overline{P}_w$).

Man erhält es bei Molekulargewichtsbestimmungen mit der Lichtzerstreuung. In gewissen Fällen ist $\overline{P}_w$ praktisch identisch mit $\overline{P}_\eta$ (siehe Abschnitt II, b und c).

Breitenbach u. Mitarb. (*23, 24*) haben gezeigt, wie in einem Zwei-Stufen-Verfahren das kinetisch wichtige Zahlenmittel aus dem mit der Lichtzerstreuung gemessenen Gewichtsmittel in guter Näherung bestimmt werden kann: Mit den ungenauen, durch Osmose bestimmten Molekulargewichten werden zunächst genäherte Werte für die Übertragungskonstante bestimmt, mit deren Hilfe dann ein Ausdruck für $\overline{P}_w/\overline{P}_n$ aufgestellt wird, welcher die durch die Übertragung bedingte Molekulargewichtsverteilung berücksichtigt. Die von Breitenbach aufgestellte Beziehung ist ein Spezialfall der in der vorliegenden Arbeit abgeleiteten Gleichungen, welche jedoch nicht mehr das Zwei-Stufen-Verfahren benötigen.

Auch Bamford u. Mitarb. (*7b*) haben einen Ansatz gemacht, das Zahlenmittel unter Beachtung der Verteilung aus der Meßgröße (in diesem Fall der Viscositätszahl) auszurechnen. Ihre Gleichung setzt jedoch die Kenntnis der Übertragungskonstanten voraus.

Zusammenfassend sei nochmals betont, daß bei den üblichen Molekulargewichtsbestimmungsmethoden keineswegs immer das für die Auswertung von Gl. I,6 notwendige Zahlenmittel erhalten wird, und daß erhebliche Unkorrektheiten durch Nichtbeachten dieser Tatsache entstehen können. Unter Berücksichtigung des Einflusses, den die verschiedenen Kettenabbruchs-Mechanismen sowie die Übertragung auf die Molekulargewichtsverteilung haben, kann man jedoch Gleichungen aufstellen, mit deren Hilfe es möglich ist, ausgehend von jedem der drei Mittelwerte des Polymerisationsgrades Übertragungskonstanten zu bestimmen.

b) Zusammenstellung aller berücksichtigten Reaktionen

Im folgenden wird eine Zusammenstellung aller in der vorliegenden Arbeit berücksichtigten Reaktionen sowie ihrer Geschwindigkeitsgleichungen gegeben. Die Bedeutung der Zeichen ergibt sich aus den Gleichungen. Alle Konzentrationen (in eckigen Klammern) werden in Mol/l gerechnet. In den Gleichungen (I,12), (I,16) und (I,17) steht X für einen beliebigen Überträger (Monomeres, Lösungsmittel, Regler, usw.).

Radikalbildung:

a) thermischer Start

$$v_{St,th} = k_{St,th}[M]^2 \qquad (I,8)$$

b) katalysierter Start

$$I \rightarrow 2R^* \qquad v_z = 2k_z[I] \qquad (I,9)$$

$$R^* + M \rightarrow RM^* \qquad v_{St} = f2k_z[I] \qquad (I,10)$$

Wachstum:

$$P_n^* + M \rightarrow P_{n+1}^* \qquad v_w = k_w [P^*] [M] \; (= v_{Br}{}^{1}) \qquad \text{(I,11)}$$

Übertragung:

$$P^* + HX \rightarrow PH + X^* \qquad v_ü = k_ü [P^*] [X] \qquad \text{(I,12)}$$

Radikalvernichtung:

a) Disproportionierung

$$P_n^* + P_m^* \rightarrow P_n + P_m \qquad v_{dis} = k_{dis} [P^*]^2 \qquad \text{(I,13)}$$

b) Kombination

$$P_n^* + P_m^* \rightarrow P_{n+m}^* \qquad v_{kom} = k_{kom} [P^*]^2 \qquad \text{(I,14)}$$

c) allgemein

$$P_n^* + P_m^* \rightarrow P_{n+m} \; (\text{oder } P_n + P_m) \qquad v_{ab} = k_{ab} [P^*]^2 \qquad \text{(I,15)}$$

d) Verzögerung

$$P^* + X^* \rightarrow PX \; (\text{oder } P + X) \qquad v_{ab'} = k_{ab'} [P^*] [X^*] \qquad \text{(I,16)}$$

$$X^* + X^* \rightarrow X_2 \; (\text{oder } X + X) \qquad v_{ab''} = k_{ab''} [X^*]^2 \qquad \text{(I,17)}$$

c) Definitionen

Bei der Ableitung der Übertragungsgleichungen werden wir uns der anschließend erläuterten Definitionen bedienen. Da sie zum Verständnis der Ableitungen unbedingt notwendig sind, möchten wir sie hier besonders herausstellen.

1. Die Polymerkette $\bar{\nu}'$. Wir bezeichnen als eine *Polymerkette* die Gesamtheit aller Monomeren, die durch eine Folge von Wachstumsschritten miteinander verbunden sind. Sie reicht vom Kettenstart oder einer Übertragung bis zum Kettenabbruch oder einer Übertragung (*130*). Ihre Kettenlänge ist gegeben durch:

$$\bar{\nu}' = \frac{v_w}{v_{ab} + \Sigma\, v_ü}\,. \qquad \text{(I,18)}$$

(In $\Sigma v_ü$ können mehrere Übertragungsreaktionen enthalten sein.)

Bei Fehlen von Übertragung ist die Polymerkette $\bar{\nu}'$ gleich der kinetischen Kettenlänge $\bar{\nu}$:

$$\bar{\nu} = \frac{v_w}{v_{ab}}\,.$$

Findet Übertragung statt, so ist $\bar{\nu}' < \bar{\nu}$.

Die Definition der Polymerkette hat sich als notwendig für die Bearbeitung der Kettenübertragung erwiesen, da mit ihrer Hilfe eine generell gültige, „allgemeine Übertragungsgleichung" aufgestellt werden kann (vgl. Gl. III,8), welche unabhängig von der Art des Kettenabbruchs sowie von der Mittelwertsbildung des verwendeten experimentellen Polymerisationsgrades ist. Aber auch für andere kinetische

---

[1] Für $P \gg 1$ ist $v_{Br} \equiv -\dfrac{d[M]}{dt} = v_w$.

Probleme hat sich die Polymerkette als ein brauchbares Hilfsmittel erwiesen (vgl. z. B. die Bestimmung des Verhältnisses der Geschwindigkeitskonstanten von Wachstum und Abbruch, Abschnitt IV, b).

2. Der Gewichtsanteil y. In allen Fällen, in welchen der Kettenabbruch ganz oder teilweise durch Kombination erfolgt, besteht das Polymere aus einem Gemisch von Molekülen, welche aus zwei Polymerketten bestehen (die durch Kombination abgebrochenen) und solchen, die aus einer Polymerkette bestehen (die durch Disproportionierung oder Übertragung abgebrochenen sowie, bei Vorliegen von Reaktion I,16, die durch diese Reaktion beendeten Moleküle).

Wir bezeichnen mit y den Gewichtsanteil derjenigen Moleküle, die aus *einer* Polymerkette bestehen.

3. Die Rechengröße $\langle \overline{P}_0 \rangle$. Wenn der Kettenabbruch ganz oder teilweise durch Kombination erfolgt, kann man eine Größe $\langle \overline{P}_0 \rangle$ definieren, die denjenigen Polymerisationsgrad angibt, der auftreten würde, wenn nur der Kombinationsabbruch allein vorläge (keine Disproportionierung, keine Übertragung). Dieser Polymerisationsgrad $\langle \overline{P}_0 \rangle$ ist eine reine Rechengröße, da bei den bisher bekannten Monomeren zumindest die Übertragung an letzterem immer stattfindet. Er kann jedoch in allen Fällen exakt berechnet werden.

4. Annahmen. Ferner liegen den kinetischen Rechnungen folgende Annahmen zugrunde:

a) Die Polymerisation erfolgt im quasistationären Zustand.

b) Polymerkette und kinetische Kettenlänge sind innerhalb eines Ansatzes als konstant anzusehen. Dies setzt voraus, daß alle Konzentrationsänderungen vernachlässigbar sind (niedrige Umsätze).

c) Molekülverzweigung kann vernachlässigt werden (sofern nicht ausdrücklich davon die Rede ist). Dies ist ebenfalls durch niedere Umsätze gewährleistet.

d) Keine Nebenreaktionen treten auf (sofern sie nicht ausdrücklich behandelt werden).

e) Die Geschwindigkeitskonstanten von Wachstum und Abbruch bzw. das Verhältnis k_w^2/k_{ab} werden als vom Medium unabhängig betrachtet.

Diese letzte Annahme bedarf vielleicht einiger Diskussion. Daß die *Abbruchs*konstante mit zunehmender Viscosität des Mediums diffusionskontrolliert wird, ist wohl ohne Zweifel [vgl. Trommsdorff (*145*), Schulz (*121*), Benson (*15*)]. Bei den hier besprochenen Messungen zur Bestimmung von Übertragungskonstanten wird jedoch meistens in verdünnter Lösung, und immer nur bis zu niederen Umsätzen ($< 5\%$) polymerisiert. Man dürfte daher im allgemeinen unterhalb der kritischen Schwelle der Diffusionskontrolle bleiben. (So findet beispielsweise Schulz (*121*) bei der Substanzpolymerisation von Methylmethacrylat bei 50° C bis zu etwa 20% Umsatz und Hayden und Melville (*54*) bei 22,5° C bis zu 10% Umsatz k_w/k_{ab} konstant.)

In schlechten Lösungsmitteln kann jedoch eine andere Störung auftreten, welche durch den Lösungszustand der Makroradikale bedingt ist: Durch Agglomeration des Polymeren, welche nicht unbedingt zu sichtbarer Heterogenität der Lösung führen muß, kann der Kettenabbruch behindert sein. Dies wurde von LIM und KOLINSKY (*73*) für Polystyrol in einer Reihe von Estern mit zunehmend schlechterer Lösefähigheit gezeigt.

Die sehr viel niedere *Wachstums*konstante ist in den hier interessierenden Systemen sicher nicht diffusionskontrolliert (*15, 121*). Dagegen wurde kürzlich von PATAT und KIRCHNER (*71, 102*) angegeben, k_w sei von der Dielektrizitätskonstante des Mediums abhängig. Eigene vergleichende Messungen der Polymerisationsgeschwindigkeit von Styrol in Lösungsmitteln mit stark variierender DK konnten diese Annahme jedoch nicht bestätigen (*59, 59a*). Falls dennoch ein irgendwie gearteter Einfluß des Mediums auf die Wachstumskonstante bestehen sollte, dürfte er wohl sehr gering sein, da es bisher nicht gelungen ist, ihn quantitativ zu fassen.

II. Die verschiedenen Mittelwerte des Polymerisationsgrades

a) Die Verteilungsfunktion der Molekulargewichte bei Vorliegen von Übertragung

Bei radikalischen Polymerisationsprozessen, bei welchen *keine* Übertragungsreaktionen auftreten, ist die Massenverteilung der Molekulargewichte nach G. V. SCHULZ (*113, 115*) gegeben durch die Gleichung:

$$m_p = \frac{(-\ln\alpha)^{k+1}}{k!} P^k \alpha^P \equiv H(P). \qquad \text{(II,1)}$$

Hierin ist m_p der Gewichtsanteil der Moleküle vom Polymerisationsgrad P; ferner ist

$$\alpha = 1 - \frac{v_{ab}}{v_w}.$$

Unter Einführen der kinetischen Kettenlänge

$$\bar{\nu} = \frac{v_w}{v_{ab}} \qquad \text{(II,2)}$$

ist α gegeben durch

$$\alpha = 1 - 1/\bar{\nu}. \qquad \text{(II,3)}$$

Der Wert k in Gl. II,1 gibt an, aus wieviel kinetischen Ketten die Polymermoleküle zusammengesetzt sind. Erfolgt der Abbruch durch Disproportionierung, ist $k = 1$; bei Kombination ist $k = 2$.

Wenn beide Abbruchmechanismen nebeneinander vorliegen, wie dies beispielsweise bei Polymethylmethacrylat (*19, 1, 131*) der Fall ist, so läßt sich die Verteilung ebenfalls nach Gl. II,1 mit stetig zwischen 1 und 2 liegenden Werten von k beschreiben. Dies kann man leicht durch Durchrechnen von Zahlenbeispielen nachprüfen (*130*).

Dieselben statistischen Überlegungen, welche zur Ableitung von Gl. II,1 führten (*113, 115*), lassen sich auch anwenden, wenn Übertragung vorliegt. Die Ketten von Monomereinheiten, die entweder einzeln (Disproportionierung und Übertragung; $k = 1$) oder kombiniert ($k = 2$) das Polymermolekül ausmachen, sind bei Vorliegen von Übertragungsreaktionen allerdings kürzer als die kinetische Kettenlänge. Während letztere durch Gl. II,2 gegeben ist, muß man hier Kettenstücke der Länge

$$\bar{\nu}' = \frac{v_w}{v_{ab} + \Sigma v_{ü}} \tag{II,4}$$

betrachten. Die Größe $\bar{\nu}'$ bezeichnen wir als *Polymerkette.*

Die Verteilung der *Polymerketten* gehorcht immer der Gleichung II,1 mit $k = 1$, wenn wir dort $\bar{\nu}'$ für P setzen, und anstelle von α die Größe

$$\alpha' = 1 - 1/\bar{\nu}' \tag{II,5}$$

einführen.

Die Verteilung der *Moleküle* ist dann

$$H(P) = \frac{(-\ln\alpha')^{k+1}}{k!} P^k \alpha'^P. \tag{II,6}$$

Bei Disproportionierungsabbruch bestehen sowohl die normal abgebrochenen, als auch die durch Übertragung abgebrochenen Moleküle aus *einer* Polymerkette, d. h. $k = 1$. Bei Konkurrenz zwischen Kombinationsabbruch einerseits und Übertragung und Disproportionierung andererseits läßt sich die Verteilung ebenfalls nach Gl. II,6 mit stetig zwischen 1 und 2 liegenden k-Werten beschreiben (vgl. das zu Gl. II,1 Gesagte).

b) Zahlen- und Gewichtsmittel des Polymerisationsgrades

Das Zahlenmittel und das Gewichtsmittel des Polymerisationsgrades sind definiert als (*113, 115*):

$$\bar{P}_n = 1/\int_0^\infty [H(P)/P]\,dP \tag{II,7}$$

und

$$\bar{P}_w = \int_0^\infty P H(P)\,dP. \tag{II,8}$$

Setzt man hierin Gl. II,6 ein und führt die Integration aus, so erhält man:

$$\bar{P}_n = -k/\ln\alpha' \tag{II,7a}$$

und

$$\bar{P}_w = -(k+1)/\ln\alpha'. \tag{II,7b}$$

Daraus folgt

$$\text{für } k = 1: \bar{P}_w/\bar{P}_n = 2 \tag{II,9}$$

$$\text{für } k = 2: \bar{P}_w/\bar{P}_n = 1{,}5 \tag{II,10}$$

und allgemein:

$$\bar{P}_w/\bar{P}_n = 1 + \frac{1}{k}.$$

Wenn man als „Uneinheitlichkeit" (*116*) die Größe

$$U \equiv \frac{\overline{P}_w}{\overline{P}_n} - 1 \tag{II,11}$$

definiert, so ist für Verteilungen nach II,6

$$U = 1/k \tag{II,12}$$

und

$$\overline{P}_w = \overline{P}_n (1 + U)\,. \tag{II,13}$$

Wie bereits erwähnt, liegen fast immer Gemische von zweikettigen ($k = 2$) und einkettigen ($k = 1$) Molekülen vor.

Für die einkettigen Moleküle ist $\overline{P}_n^{\mathrm{I}} = \overline{\nu}'$, und wegen (II,9), $\overline{P}_w^{\mathrm{I}} = 2\overline{\nu}'$.

Für die zweikettigen ist $\overline{P}_n^{\mathrm{II}} = 2\overline{\nu}'$, und, wegen (II,10), $\overline{P}_w^{\mathrm{II}} = 3\overline{\nu}'$.

Bezeichnet man mit y den *Gewichtsanteil der einkettigen Moleküle*, so lassen sich für ein solches Gemisch folgende Gleichungen für das Zahlenmittel, das Gewichtsmittel, und nach (II,11), für die Uneinheitlichkeit angeben:

$$\overline{P}_n = \frac{1}{y/\overline{P}_n^{\mathrm{I}} + (1-y)/\overline{P}_n^{\mathrm{II}}} = \frac{1}{y/\overline{\nu}' + (1-y)/2\overline{\nu}'}$$

$$\overline{P}_n = \frac{2\overline{\nu}'}{y+1} \tag{II,14}$$

$$\overline{P}_w = y\,\overline{P}_w^{\mathrm{I}} + (1-y)\,\overline{P}_w^{\mathrm{II}} = y\,2\overline{\nu}' + (1-y)\,3\overline{\nu}'$$

$$\overline{P}_w = \overline{\nu}'(3-y) \tag{II,15}$$

$$U = 0{,}5 + y(1 - 0{,}5y)\,. \tag{II,16}$$

c) Viscositätsmittel des Polymerisationsgrades

1. Definition des Viscositätsmittels. Wie bereits in Abschnitt I a erwähnt, hat MEYERHOFF (*88*) gezeigt, daß die Bestimmung des Polymerisationsgrades mit der Ultrazentrifuge (Geschwindigkeitstyp) in Verbindung mit der Diffusion einen Mittelwert liefert, der praktisch der Mittelung durch die Viscosität entspricht. MEYERHOFF fand nämlich bei der Untersuchung von Polymethacrylaten sehr verschiedene Einheitlichkeit, daß innerhalb gewisser Grenzen[1] unabhängig von der Molekulargewichtsverteilung des einzelnen Präparates ein linearer Zusammenhang zwischen dem Logarithmus der Viscositätszahl $[\eta]$ und dem Logarithmus des Polymerisationsgrades besteht. Das bedeutet aber, daß die Beziehung

$$[\eta] = K \cdot P^{\alpha} \tag{II,17}$$

[1] Präparate, bei denen absichtlich Verteilungen angestrebt wurden, die wesentlich breiter sind als die Normalverteilung nach Gl. II,1 bzw. II,6, gehorchten nicht der Beziehung. Dasselbe gilt für verzweigte Moleküle (*92*).

für die einzelnen Komponenten einer polymolekularen Substanz mit denselben Zahlenwerten für K und α gilt, wie für die Gesamtsubstanz. Das heißt

$$[\eta]_i = K_i \cdot P_i^{a_i} \tag{II,17a}$$

mit $K_i = K$ und $a_i = a$.

G. V. Schulz (*118*) hat gezeigt, daß unter dieser Voraussetzung der resultierende Mittelwert gut definiert und durch folgende Gleichung gegeben ist:

$$\overline{P}_\eta = \left(\int_0^\infty P^a H(P)\, d\,P \right)^{1/a}. \tag{II,18}$$

$H(P)$ ist wieder die Massenverteilungsfunktion nach II,6, a ist der Exponent in Gl. II,17. Für $a = 1$ ist $\overline{P}_\eta$ identisch mit $\overline{P}_w$.

Bei Eichung der Viscosität mit der Ultrazentrifuge ist also jedem $[\eta]$ eindeutig ein Polymerisationsgrad zuzuordnen, welcher durch Gl. II,18 definiert ist.

Falls für eine polymerhomopologe Reihe Gl. II,17 *nicht* gilt (Krümmung in der doppelt logarithmischen Darstellung von $[\eta]$ gegen P), dann sind die Voraussetzungen $K_i = K$ und $a_i = a$ nicht gegeben, und der Mittelwert ist nicht durch Gl. II,18 definiert. Meist wird es sich in diesen Fällen um verzweigte Polymere handeln [z. B. Polyäthylen (*144*)], und dann ist eine viscosimetrische Molekulargewichtsbestimmung sowieso nicht angebracht (*92*). Man muß in diesen Fällen auf die experimentelle Einfachheit von $\overline{P}_\eta$ verzichten und Absolutmethoden heranziehen.

2. Ableitung einer Näherungsgleichung für P_η (*130*). Wenn man in Gl. II,6 den Ausdruck $-k/\ln \alpha'$ durch $\overline{P}_n$ ersetzt (siehe Gl. II,7a), so erhält man

$$H(P) = \frac{k}{\overline{P}_n\, k!} \left(\frac{k\,P}{\overline{P}_n} \right)^k e^{-kP/\overline{P}_n}.$$

Dieses eingesetzt in (II,18) ergibt:

$$\overline{P}_\eta = \left[\int_0^\infty \frac{k\,P^a}{\overline{P}_n\, k!} \left(\frac{k\,P}{\overline{P}_n} \right)^k e^{-kP/\overline{P}_n}\, d\,P \right]^{1/a}. \tag{II,19}$$

Die Lösung dieses Integrals liefert:

$$\overline{P}_\eta = \overline{P}_n \frac{1}{k} \left[\frac{(k+a)!}{k!} \right]^{1/a},$$

bzw., da man k und a als stetige Größen betrachten kann:

$$\overline{P}_\eta = \overline{P}_n \frac{1}{k} \left[\frac{\Gamma(k+a+1)}{\Gamma(k+1)} \right]^{1/a}. \tag{II,20}$$

Für den hier in Frage kommenden Bereich kann man statt dessen die bequemer zu handhabende Näherungsgleichung

$$\overline{P}_\eta \approx \overline{P}_n \left(1 + \frac{1+a}{2k}\right) \tag{II,20a}$$

verwenden[1]. Setzen wir darin

$$b \equiv \frac{1+a}{2}, \tag{II,21}$$

so erhalten wir

$$\overline{P}_\eta = \overline{P}_n \left(1 + \frac{b}{k}\right)$$

und, unter Verwendung von (II,12),

$$\overline{P}_\eta = \overline{P}_n (1 + bU)\,. \tag{II,22}$$

Diese Gleichung erlaubt uns nun, $\overline{P}_\eta$ auch als Funktion von y und $\bar{\nu}'$ anzugeben, ebenso wie das in Gl. II,14 und 15 für $\overline{P}_n$ und $\overline{P}_w$ geschehen ist.

Aus (II,22), (II,14) und (II,16) folgt nämlich:

$$\overline{P}_\eta = \bar{\nu}' \frac{2 + b + 2by - by^2}{y+1}\,. \tag{II,23}$$

III. Formale Beziehungen

a) Ableitung der allgemeinen Übertragungsgleichung

Aus den Definitionen für die kinetische Kettenlänge (vgl. Gl. I,19)

$$\bar{\nu} = \frac{v_w}{v_{ab}} \tag{III,1}$$

und für die Polymerkette (vgl. I,18)

$$\bar{\nu}' = \frac{v_w}{v_{ab} + \Sigma v_ü} \tag{III,2}$$

ergibt sich:

$$\frac{1}{\bar{\nu}'} - \frac{1}{\bar{\nu}} = \Sigma \frac{v_ü}{v_w}\,. \tag{III,3}$$

Wenn wir in (III,1) die kinetischen Ausdrücke für die Geschwindigkeiten von Wachstum und Abbruch:

$$v_w = k_w [P^*]\,[M] \tag{III,4}$$

und

$$v_{ab} = k_{ab} [P^*]^2 \tag{III,5}$$

---

[1] Man kann für jeden Einzelfall auf Gl. II,20 zurückgreifen und die Güte der Näherung überprüfen.

einsetzen und ferner berücksichtigen, daß für $P \gg 1$

$$-\frac{d[M]}{dt} = v_{Br} = v_w \,, \tag{III,6}$$

so erhalten wir

$$\bar{\nu} = \frac{k_w^2 [M]^2}{k_{ab}\, v_{Br}} \,. \tag{III,7}$$

Wenn man dies in Gl. III,3 einsetzt, so erhält man die grundlegende, ganz allgemein gültige Gleichung für alle Übertragungsvorgänge:

$$\frac{1}{\bar{\nu}'} - \frac{k_{ab}\, v_{Br}}{k_w^2 [M]^2} = \Sigma \frac{v_{ü}}{v_w} \,. \tag{III,8}$$

Um diese Gleichung auswerten zu können, muß man zunächst die Polymerkette $\bar{\nu}'$ durch den gemessenen Mittelwert des Polymerisationsgrades, $\bar{P}_n$, $\bar{P}_w$ oder $\bar{P}_\eta$, ersetzen. Dies geschieht nach den bereits abgeleiteten Gleichungen (vgl. II,14, II,15 und II,23):

$$\bar{\nu}' = \bar{P}_n \frac{y+1}{2} \tag{III,9}$$

$$\bar{\nu}' = \bar{P}_w \frac{1}{3-y} \tag{III,10}$$

$$\bar{\nu}' = \bar{P}_\eta \frac{y+1}{2+b+2by-by^2} \,. \tag{III,11}$$

Bei der Berechnung des Gewichtsbruches y muß nun in jedem Einzelfall berücksichtigt werden, welche Reaktionen zu Molekülen führen, die nur aus einer Polymerkette bestehen, und welche somit den Bruchteil y ausmachen. Bei Disproportionierungsabbruch z. B. ist in jedem Fall $y = 1$, und die Übertragungsgleichungen werden sehr einfach (Abschnitt III, b). Bei gleichzeitigem Vorliegen von Kombination und Disproportionierung gehören sowohl die durch Übertragung, als auch die durch Disproportionierung abgebrochenen Moleküle zum Bruchteil y. Kenntnis des Verhältnisses der beiden Abbruchsreaktionen ist daher notwendig (Abschnitt III, d). Bei Verzögerung muß beachtet werden, daß auch durch Abbruch von Polymerradikalen durch Überträgerradikale Moleküle mit nur einer Polymerkette entstehen können (Abschnitt III, e).

Ferner ist es, wie man aus Gl. III,8 ersehen kann, unerläßlich, in jedem Fall auch die Bruttogeschwindigkeit zu bestimmen. Außerdem muß das Verhältnis der Geschwindigkeitskonstanten von Wachstum und Abbruch, k_w^2/k_{ab}, bekannt sein. (Über die Bestimmung dieser aus kinetischen Daten leicht berechenbaren Größe siehe Abschnitt IV, b.)

Die Summe der Übertragungsterme rechts in Gl. III,8 setzt sich aus so vielen Summanden zusammen, wie Überträger-Substanzen im System vorliegen. Durch Einsetzen der entsprechenden Geschwindigkeitsgleichungen erhält man:

$$\Sigma \frac{v_ü}{v_w} = C_M + C_{LM} \frac{[LM]}{[M]} + C_I \frac{[I]}{[M]} + C_X \frac{[X]}{[M]} + \cdots \quad \text{(III,12)}$$

$C_M, C_{LM}, C_I, C_X, \ldots$ sind die relativen Übertragungskonstanten ($C \equiv k_ü/k_w$) des Monomeren, des Lösungsmittels, des Initiators, eines beliebigen Überträgers X, usw.; in eckigen Klammern stehen die Konzentrationen in Mol/l. Zur Bestimmung der einzelnen Übertragungskonstanten sind im allgemeinen Versuchsserien nötig. So liefert z. B. bei der thermischen Polymerisation in Lösung die graphische Darstellung von $\Sigma(v_ü/v_w)$ als Funktion von $[LM]/[M]$ als Ordinatenabschnitt C_M, als Steigung C_{LM}. Darstellung von $\Sigma(v_ü/v_w)$ als Funktion von $[I]/[M]$ bei konstant gehaltener Lösungsmittelkonzentration liefert als Ordinatenabschnitt $C_M + C_{LM} \frac{[LM]}{[M]}$, als Steigung C_I, usw. (vgl. Gl. III,12).

Beispiele für die Auswertung werden in Abschnitt V gegeben.

b) Kettenübertragung bei Abbruch durch Disproportionierung

In der allgemeinen Übertragungsgleichung III,8 wird die Polymerkette durch die experimentellen Mittelwerte des Polymerisationsgrades ersetzt nach Gl. III,9—11. Da bei Abbruch durch Disproportionierung alle Moleküle aus *einer* Polymerkette

$$\bar{\nu}' = \frac{v_w}{v_{ab} + \Sigma v_ü}$$

bestehen, ist immer $y = 1$. Die Übertragungsgleichungen für die verschiedenen Mittelwerte lassen sich daher in einfacher Weise ableiten.

Das *Zahlenmittel* ist nach (III,9):

$$\bar{P}_n = \bar{\nu}' .$$

Somit folgt aus (III,8):

$$\frac{1}{\bar{P}_n} - \frac{k_{ab}\, v_{Br}}{k_w^2\, [M]^2} = \Sigma \frac{v_ü}{v_w} . \quad \text{(III,13)}$$

Das *Gewichtsmittel* ist nach (III,10):

$$\bar{P}_w = 2\bar{\nu}' .$$

Dies eingesetzt in (III,8) ergibt:

$$\frac{2}{\bar{P}_w} - \frac{k_{ab}\, v_{Br}}{k_w^2\, [M]^2} = \Sigma \frac{v_ü}{v_w} . \quad \text{(III,14)}$$

Das *Viscositätsmittel* ist nach (III,11):

$$\bar{P}_\eta = (1 + b)\bar{\nu}' .$$

Einsetzen in (III,8) liefert:

$$\frac{1+b}{\overline{P}_\eta} - \frac{k_{ab}\, v_{Br}}{k_w^2\,[M]^2} = \Sigma \frac{v_{ü}}{v_w}. \qquad \text{(III,15)}$$

c) Kettenübertragung bei Abbruch durch Kombination

Der Ausgangspunkt ist wiederum die allgemeine Übertragungsgleichung III,8. Das Ersetzen der Polymerkette $\bar{\nu}'$ durch die experimentellen Mittelwerte des Polymerisationsgrades nach Gl. III,9—11 erweist sich jedoch als nicht mehr so einfach wie im vorigen Abschnitt. Da bei Abbruch durch Kombination Mischungen von kombinierten Molekülen ($k = 2$) und übertragenen Molekülen ($k = 1$) vorliegen, ist y nicht mehr 1, sondern von den jeweiligen Versuchsbedingungen abhängig. Der Gewichtsanteil y muß daher als Funktion meßbarer Größen ausgedrückt und für jeden Einzelversuch ausgerechnet werden.

Der Gewichtsanteil y ist durch die Geschwindigkeiten von Übertragungs- und Abbruchreaktionen bestimmt.

$$y = \frac{\Sigma v_{ü}}{v_{ab} + \Sigma v_{ü}}; \qquad 1 - y = \frac{v_{ab}}{v_{ab} + \Sigma v_{ü}}.$$

Berücksichtigung von (III,1) und (III,2) gibt dann:

$$y = 1 - \frac{\bar{\nu}'}{\bar{\nu}}. \qquad \text{(III,16)}$$

Auch bei dieser Gleichung können wir die Polymerkette $\bar{\nu}'$ durch den experimentellen Polymerisationsgrad nach (III,9—11) ersetzen.

Die kinetische Kettenlänge $\bar{\nu}$ ersetzen wir nun durch jenen Polymerisationsgrad $\langle \overline{P}_0 \rangle$, der bei völliger Abwesenheit von Kettenübertragung auftreten würde (vgl. Abschnitt I, c 3).

In einem solchen Polymeren wären alle Moleküle durch Kombination abgebrochen. Es wäre daher $y = 0$, und die Polymerkette wäre gleich der kinetischen Kettenlänge (vgl. Gl. III,1 und III,2). Aus (III,9—11) folgt dann:

$$\langle \overline{P}_{n,0} \rangle = 2\bar{\nu} \qquad \text{(III,17)}$$

$$\langle \overline{P}_{w,0} \rangle = 3\bar{\nu} \qquad \text{(III,18)}$$

$$\langle \overline{P}_{\eta,0} \rangle = (2+b)\,\bar{\nu}. \qquad \text{(III,19)}$$

Ein Zusammenhang zwischen dem Gewichtsbruch y und der Molekulargewichtsverminderung durch Übertragung $\overline{P}/\langle \overline{P}_0 \rangle$ kann nun ohne Schwierigkeit formuliert werden. So folgt aus (III,16), (III,9) und (III,17):

$$y = \frac{1 - \overline{P}_n/\langle \overline{P}_{n,0} \rangle}{1 + \overline{P}_n/\langle P_{n,0} \rangle}, \qquad \text{(III,20)}$$

aus (III,16), (III,10) und (III,18):

$$y = 2 \overset{(+)}{-} \sqrt{1 + 3\overline{P}_w/\langle\overline{P}_{w,0}\rangle} \qquad \text{(III,21)}$$

und schließlich aus (III,16), (III,11) und (III,19):

$$\frac{(1-y)\,(2+b+2by-by^2)}{(1+y)\,(2+b)} = \frac{\overline{P}_\eta}{\langle\overline{P}_{\eta,0}\rangle}. \qquad \text{(III,22)}$$

Die letzte Gleichung läßt sich zwar nicht so einfach nach y auflösen wie die beiden vorangegangenen. Man kann sie jedoch graphisch lösen, indem man $\overline{P}_\eta/\langle\overline{P}_{\eta,0}\rangle$ als Funktion von y aufträgt und dabei b als Parameter verwendet, bzw. dieses sogleich nach (II,21) durch a ersetzt. Tab. 1 gibt die daraus durch Interpolation erhaltenen Wertepaare $[\overline{P}_\eta/\langle\overline{P}_{\eta,0}\rangle; y]$ für verschiedene Werte von a in dem praktisch vorkommenden Bereich $0{,}5 \gtreqless a \gtreqless 1$ an.

Tabelle 1. *y in Abhängigkeit von $\overline{P}_\eta/\langle\overline{P}_{\eta,0}\rangle$ und a, nach Gl. III,22 interpoliert; $b = (1+a)/2$ nach (II,21)*

$\overline{P}_\eta/\langle\overline{P}_{\eta,0}\rangle$ \ a	0,5	0,6	0,7	0,8	0,9	1,0
1,0	0	0	0	0	0	0
0,9	0,068	0,070	0,072	0,074	0,076	0,078
0,8	0,146	0,149	0,152	0,155	0,157	0,159
0,7	0,226	0,229	0,232	0,235	0,238	0,240
0,6	0,308	0,311	0,314	0,317	0,320	0,323
0,5	0,404	0,407	0,410	0,413	0,416	0,419
0,4	0,501	0,504	0,507	0,510	0,513	0,516
0,3	0,607	0,610	0,613	0,616	0,619	0,622
0,2	0,725	0,728	0,731	0,734	0,736	0,738
0,1	0,854	0,856	0,858	0,859	0,860	0,861

Aus Tab. 1 ist zu entnehmen, daß y für konstant gehaltenes $\overline{P}_\eta/\langle\overline{P}_{\eta,0}\rangle$ sich nur sehr wenig ändert, wenn der Exponent a aus Gl. II,17 von 0,5—1,0 variiert. In $\langle\overline{P}_{\eta,0}\rangle$ ist a jedoch ebenfalls enthalten. Es treten daher, wie wir weiter unten sehen werden, systematische Fehler auf, wenn der Unterschied zwischen $\overline{P}_\eta$ und $\overline{P}_w$ vernachlässigt wird.

Nun sind alle Grundlagen gegeben, um, ausgehend von der allgemeinen Übertragungsgleichung III,8, die Gleichungen für die drei verschiedenen Mittelwerte des Polymerisationsgrades bei Kombinationsabbruch aufzustellen.

1. Zahlenmittel

Aus (III,8) und (III,9) folgt:

$$\frac{2}{\overline{P}_n(y+1)} - \frac{k_{ab}\,v_{Br}}{k_w^2\,[M]^2} = \Sigma\frac{v_{ü}}{v_w}.$$

Wir schreiben diese Gleichung in der Form:

$$\frac{A_n}{\overline{P}_n} - \frac{k_{ab}\, v_{Br}}{k_w^2\,[M]^2} = \Sigma \frac{v_{ü}}{v_w} \qquad \text{(III,23)}$$

mit

$$A_n = \frac{2}{y+1}.$$

Mit Hilfe von III,20 kann A_n direkt als Funktion von $\overline{P}_n/\langle \overline{P}_{n,0}\rangle$ angegeben werden:

$$A_n = 1 + \frac{\overline{P}_n}{\langle \overline{P}_{n,0}\rangle} \qquad \text{(III,24)}$$

$\langle \overline{P}_{n,0}\rangle$ wird hierzu berechnet nach der sich aus (III,17) und (III,7) ergebenden Gleichung:

$$\langle \overline{P}_{n,0}\rangle = \frac{2\,k_w^2\,[M]^2}{k_{ab}\, v_{Br}}. \qquad \text{(III,25)}$$

2. Gewichtsmittel

Die Aufstellung der Gleichung für das Gewichtsmittel erfolgt analog derjenigen für das Zahlenmittel:

Aus (III,8) und (III,10) folgt

$$\frac{A_w}{\overline{P}_w} - \frac{k_{ab}\, v_{Br}}{k_w^2\,[M]^2} = \Sigma \frac{v_{ü}}{v_w} \qquad \text{(III,26)}$$

mit

$$A_w = 3 - y. \qquad \text{(III,27a)}$$

Unter Verwendung von (III,21) drücken wir A_w als Funktion von $\overline{P}_w/\langle \overline{P}_{w,0}\rangle$ aus:

$$A_w = 1 + \sqrt{1 + 3\,\overline{P}_w/\langle \overline{P}_{w,0}\rangle}, \qquad \text{(III,27b)}$$

wobei $\langle \overline{P}_{w,0}\rangle$ gegeben ist durch die aus (III,18) und (III,7) folgende Gleichung:

$$\langle \overline{P}_{w,0}\rangle = \frac{3\,k_w^2\,[M]^2}{k_{ab}\, v_{Br}}. \qquad \text{(III,28)}$$

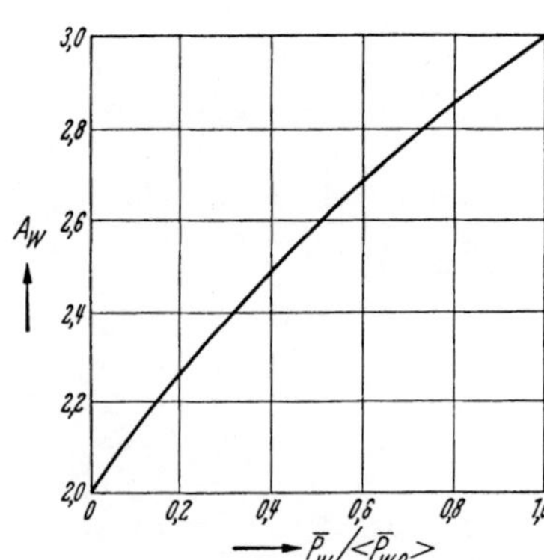

Abb. 1. Der Faktor A_w (zur Auswertung von Gl. III,26) in Abhängigkeit von $\overline{P}_w/\langle \overline{P}_{w,0}\rangle$ nach Gl. III,27b

In Abb. 1 ist A_w als Funktion von $\overline{P}_w/\langle \overline{P}_{w,0}\rangle$ graphisch dargestellt. Anstatt A_w jedesmal auszurechnen, kann man den Wert auch aus der Kurve entnehmen.

3. Viscositätsmittel

Wir gehen wieder in gleicher Weise vor, wie bei den beiden vorangegangenen Fällen:

Aus (III,8) und (III,11) folgt

$$\frac{A_\eta}{\overline{P}_\eta} - \frac{k_{ab}\, v_{Br}}{k_w^2\,[M]^2} = \Sigma \frac{v_{ü}}{v_w} \qquad \text{(III,29)}$$

mit

$$A_\eta = \frac{2 + b + 2by - by^2}{y + 1} \qquad \text{(III,30)}$$

Man kann y aus Tab. 1 entnehmen (bzw. interpolieren), wozu man zunächst $\langle \overline{P}_{\eta,0} \rangle$ berechnen muß, nach der aus (III,19) und (III,7) folgenden Gleichung:

$$\langle \overline{P}_{\eta,0} \rangle = (2 + b) \frac{k_w^2 [M]^2}{k_{ab} v_{Br}}, \qquad \text{(III,31)}$$

und ferner berücksichtigen muß, daß $b = (1 + a)/2$ (siehe II,21).

Zur Vereinfachung der Auswertung ist der Faktor A_η in Abb. 2 direkt als Funktion von $\overline{P}_\eta / \langle \overline{P}_{\eta,0} \rangle$ dargestellt, wobei a als Parameter verwendet wurde.

Für $a = b = 1$ ist $A_\eta = A_w$ (vgl. III,30 und III,27a). Die oberste Kurve in Abb. 2 ist daher identisch mit derjenigen in Abb. 1.

Der Faktor A_η ist nun deutlich von a abhängig. Der Fehler, den man macht, wenn man den Unterschied zwischen $\overline{P}_\eta$ und $\overline{P}_w$ vernachlässigt, beträgt im Falle des Styrols ($a = 0{,}72$) etwa 6—7% in A_η. (Hierbei ist berücksichtigt, daß a in der Abscisse in Abb. 2 auf dem Wege über $\langle \overline{P}_{\eta,0} \rangle$ ebenfalls enthalten ist.) Für kleine Übertragungseffekte, bei welchen die beiden Terme links in Gl. III,29 von derselben Größenordnung sind, ist der Fehler in $\Sigma(v_ü/v_w)$ jedoch ungleich größer. Da die Gleichungen für $\overline{P}_\eta$ ebenso einfach zu handhaben sind, wie diejenigen für $\overline{P}_w$, ist es deshalb unbedingt zu empfehlen, diese Fehlerquelle auszuscheiden.

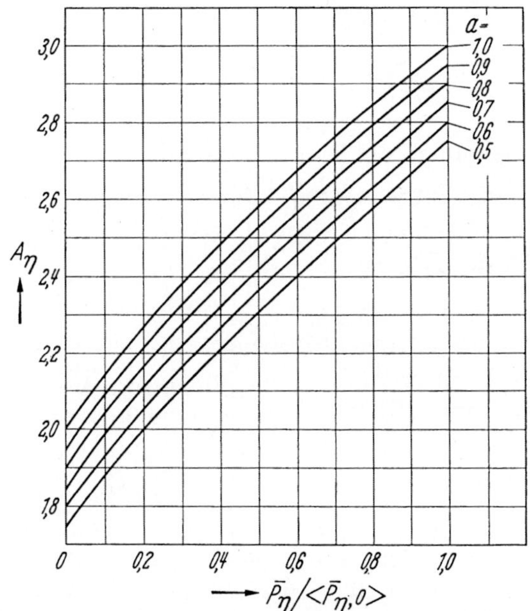

Abb. 2. Der Faktor A_η (zur Auswertung von Gl. III,29) in Abhängigkeit von $\overline{P}_\eta / \langle \overline{P}_{\eta,0} \rangle$ nach Gl. III,30, in Verbindung mit Tab. 1

d) Kettenübertragung bei gleichzeitigem Vorliegen von Disproportionierungs- und Kombinationsabbruch

Die Ableitung der Gleichungen für diesen Fall, der z. B. bei Polymethylmethacrylat verwirklicht ist (*19, 1, 131*), erfolgt analog jener bei reinem Kombinationsabbruch. Es muß lediglich berücksichtigt werden, daß die durch Disproportionierung abgebrochenen Polymerketten mit in dem Gewichtsbruch y enthalten sind und nur die durch Kombination abgebrochenen Ketten den Bruchteil $(1-y)$ ausmachen.

Die kinetische Kettenlänge ist in diesem Fall gegeben durch:

$$\bar{\nu} = \frac{v_w}{v_{dis} + v_{com}},$$

die Polymerkette ist entsprechend:

$$\bar{\nu}' = \frac{v_w}{v_{dis} + v_{com} + \Sigma v_{ü}}. \qquad \text{(III,32)}$$

v_{dis} und v_{com} sind die Geschwindigkeiten der beiden Abbruchsreaktionen, Disproportionierung und Kombination:

$$v_{dis} = k_{dis}[P^*]^2$$
$$v_{com} = k_{com}[P^*]^2.$$

Die Summe der beiden Konstanten k_{dis} und k_{com} ergibt sich bei der üblichen Auswertung kinetischer Messungen (vgl. Abschnitt IV, a):

$$k_{dis} + k_{com} = k_{ab}.$$

Es ist daher auch

$$v_{dis} + v_{com} = v_{ab}. \qquad \text{(III,33)}$$

Daraus folgt, daß die kinetische Kettenlänge hier, wie in den vorangegangenen Abschnitten, durch Gl. III,7 gegeben ist. Das heißt, die allgemeine Übertragungsgleichung (vgl. III,8):

$$\frac{1}{\bar{\nu}'} - \frac{k_{ab}\, v_{Br}}{k_w^2\,[M]^2} = \Sigma \frac{v_{ü}}{v_w}$$

ist auch in diesem Fall anwendbar. Das Ersetzen von $\bar{\nu}'$ durch die experimentellen Mittelwerte des Polymerisationsgrades geschieht auch hier nach den Gleichungen III,9—11.

Der Gewichtsanteil y der aus *einer* Polymerkette bestehenden Moleküle ist nun aber gegeben durch

$$y = \frac{v_{dis} + \Sigma v_{ü}}{v_{dis} + v_{com} + \Sigma v_{ü}},$$

und daher ist:

$$1 - y = \frac{v_{com}}{v_{dis} + v_{com} + \Sigma v_{ü}}. \qquad \text{(III,34)}$$

Wir definieren nun eine Größe $\langle\bar{\nu}\rangle$ als diejenige kinetische Kettenlänge, die auftreten würde, wenn nur der Kombinationsabbruch vorläge. Dies ist wieder eine reine Rechengröße, jedoch exakt angebbar, nämlich:

$$\langle\bar{\nu}\rangle = \frac{v_w}{v_{com}}. \qquad \text{(III,35)}$$

Einsetzen von (III,32) und (III,35) in (III,34) gibt die der Gleichung III,16 entsprechende Beziehung:

$$y = 1 - \frac{\bar{\nu}'}{\langle\bar{\nu}\rangle}. \tag{III,36}$$

Hierin ersetzen wir wieder $\bar{\nu}'$ durch den experimentellen Polymerisationsgrad nach III,9—11 und $\langle\bar{\nu}\rangle$ durch denjenigen Polymerisationsgrad $\langle\bar{P}_0\rangle$, der auftreten würde, wenn keine Übertragung *und nur der Kombinationsabbruch* vorläge. Für diesen ist $k = 2$, $y = 0$, und wir erhalten (vgl. III,17—19):

$$\langle\bar{P}_{n,0}\rangle = 2\langle\bar{\nu}\rangle \tag{III,37}$$

$$\langle\bar{P}_{w,0}\rangle = 3\langle\bar{\nu}\rangle \tag{III,38}$$

$$\langle\bar{P}_{\eta,0}\rangle = (2 + b)\langle\bar{\nu}\rangle \tag{III,39}$$

Das Ersetzen von $\bar{\nu}'$ und $\langle\bar{\nu}\rangle$ in Gl. III,36 in der besprochenen Weise führt exakt zu den Gleichungen III,20—22. Das bedeutet aber, daß man die im vorangegangenen Abschnitt für reinen Kombinationsabbruch abgeleiteten Gleichungen auch in diesem Fall anwenden kann, sofern man nur die $\langle\bar{P}_0\rangle$ nach III,37—39 (anstelle von III,17—19) berechnet.

Dazu muß jedoch bekannt sein, welchen Anteil vom gesamten Kettenabbruch der Kombinationsabbruch ausmacht (vgl. III,35).

Wir definieren δ als das Verhältnis der Geschwindigkeiten der beiden Abbruchsreaktionen:

$$\delta \equiv \frac{v_{dis}}{v_{com}}. \tag{III,40}$$

Aus dieser Definition folgt, bei Berücksichtigung von (III,33):

$$v_{com} = \frac{v_{ab}}{\delta + 1}. \tag{III,41}$$

Einsetzen von III,41 in III,35 liefert, bei Beachten von III,4—6, schließlich:

$$\langle\bar{\nu}\rangle = \frac{k_w^2\,[M]^2}{k_{ab}\,v_{Br}}(\delta + 1). \tag{III,42}$$

Das Verhältnis δ der beiden Abbruchskonstanten ist mit Hilfe radioaktiver Endgruppenbestimmungen berechenbar und ist z. B. für Polymethylmethacrylat in Abhängigkeit von der Temperatur bekannt (siehe Abschnitt IV, a).

Im einzelnen erfolgt die Berechnung von $\Sigma v_{ü}/v_w$ bei gleichzeitigem Vorliegen von Kombinations- und Disproportionierungsabbruch somit nach folgenden Gleichungen:

1. Zahlenmittel

$$\frac{A_n}{\bar{P}_n} - \frac{k_{ab}\,v_{Br}}{k_w^2\,[M]^2} = \Sigma\frac{v_{ü}}{v_w} \qquad \text{Vgl. (III,23)}$$

mit

$$A_n = 1 + \frac{\overline{P}_n}{\langle \overline{P}_{n,0} \rangle} . \qquad \text{Vgl. (III,24)}$$

Aus (III,37) und (III,42) folgt:

$$\langle \overline{P}_{n,0} \rangle = \frac{2 k_w^2 [M]^2}{k_{ab} v_{Br}} (\delta + 1) .$$

2. *Gewichtsmittel*

$$\frac{A_w}{\overline{P}_w} - \frac{k_{ab} v_{Br}}{k_w^2 [M]^2} = \Sigma \frac{v_ü}{v_w} \qquad \text{Vgl. (III,26)}$$

mit

$$A_w = 1 + \sqrt{1 + 3 \overline{P}_w / \langle \overline{P}_{w,0} \rangle} . \qquad \text{Vgl. (III,27b)}$$

Aus (III,38) und (III,42) ergibt sich:

$$\langle \overline{P}_{w,0} \rangle = \frac{3 k_w^2 [M]^2}{k_{ab} v_{Br}} (\delta + 1) .$$

3. *Viscositätsmittel*

$$\frac{A_\eta}{\overline{P}_\eta} - \frac{k_{ab} v_{Br}}{k_w^2 [M]^2} = \Sigma \frac{v_ü}{v_w} . \qquad \text{Vgl. (III,29)}$$

A_η wird aus Abb. 2 entnommen, wobei (siehe (III,39) und (III,42)):

$$\langle \overline{P}_{\eta,0} \rangle = (2 + b) \frac{k_w^2 [M]^2}{k_{ab} v_{Br}} (\delta + 1) .$$

e) Bestimmung der Übertragungskonstante bei Vorliegen von Verzögerung

1. Ursache der Verzögerung und bisherige Ansätze zu ihrer Berücksichtigung. Wenn durch den Zusatz eines Kettenüberträgers (X) die Bruttogeschwindigkeit der Polymerisation vermindert wird, so ist das ein Zeichen dafür, daß die bei der Übertragungsreaktion

$$P^* + HX \rightarrow PH + X^* \qquad \text{(III,43)}$$

entstehenden Überträgerradikale (X^*) stärker als die wachsenden Ketten zu Abbruchsreaktionen neigen.

Die Überträgerradikale können reagieren nach

$$X^* + M \xrightarrow{k_w'} XM^* \quad (v_w') \qquad \text{(III,44)}$$

$$X^* + P^* \xrightarrow{k_{ab}'} XP \quad (v_{ab}') \qquad \text{(III,45)}$$

$$X^* + X^* \xrightarrow{k_{ab}''} XX \quad (v_{ab}'') . \qquad \text{(III,46)}$$

Verzögerung kann allerdings auch dadurch zustande kommen, daß das abbrechende Radikal durch Kopolymerisation mit der Substanz X entsteht:

$$P^* + X \rightarrow PX^* . \qquad \text{(III,47)}$$

Auch hierdurch wird der Polymerisationsgrad herabgesetzt, und man kann diese Verminderung fälschlicherweise als eine Übertragungsreaktion deuten. Reaktionskinetisch ist zwischen den beiden Möglichkeiten nicht zu unterscheiden. Hier soll jedoch nur der Fall behandelt werden, daß eindeutige Übertragung vorliegt. Es soll aber darauf hingewiesen werden, daß Übertragungskonstanten von Substanzen mit polymerisierbaren Doppelbindungen aus diesem Grunde immer zweifelhaft sind (siehe z. B. Abschnitt V, a 2).

Zur Berücksichtigung der Verzögerung bei der Bestimmung von Übertragungskonstanten sind bereits verschiedene Ansätze gemacht worden. Bei einigen von ihnen (z. B. MERRET u. Mitarb. (*2*) und JENKINS (*66*) wurde die Voraussetzung gemacht, daß die Geschwindigkeitskonstanten des Abbruches zwischen ungleichen Radikalen gleich dem geometrischen Mittel der beiden Konstanten zwischen jeweils gleichen Radikalen ist (geometric mean assumption). Diese Annahme ist jedoch durch zahlreiche Kopolymerisationsversuche von MAYO und WALLING (*84, 151*) und MELVILLE und VALENTINE (*86, 87*) widerlegt, wobei sich ergeben hat, daß die Konstante des Abbruches zwischen ungleichen Radikalen praktisch immer größer ist als die beiden anderen.

Von BURNETT und LOAN (*29*) wurden Gleichungen abgeleitet, die jedoch nur für sehr schwache Verzögerung anwendbar sind, da sie Reaktion III,45 vernachlässigen.

Ein von JENKINS (*66*) ausgearbeitetes Gleichungssystem liefert im Falle von starken Verzögerungseffekten einen Näherungswert für die Übertragungskonstante, versagt jedoch bei geringer Verzögerung. Die Ursache hierfür ist, daß die Startgeschwindigkeit konstant eingesetzt wird. Diese ist aber bekanntlich, auf dem Wege über die Radikalausbeute des Initiators, von der Monomerkonzentration abhängig (siehe S. 522). Sobald die beiden Effekte (Verzögerung durch Gl. III,45 und III,46 bzw. Verminderung der Geschwindigkeit durch Verringerung der Radikalausbeute) von derselben Größenordnung sind, versagt die Methode. In jedem Fall erhält JENKINS nur eine obere und untere Grenze für die Übertragungskonstante, je nachdem, ob er die bereits erwähnte Annahme des geometrischen Mittels für die Konstante k_{ab}' macht oder aber $k_{ab}'' = 0$ setzt.

Ferner wurden von KICE (*70*) Gleichungen für die verzögerte Polymerisation gegeben. Diese sind praktisch voraussetzungslos und scheinen zur Bearbeitung des Problems gut geeignet. Da es KICE in seiner Arbeit primär auf den *Verzögerungs*effekt[1] ankam, verwendet er als Meßgröße nur die Bruttogeschwindigkeit. (Letzteres ist übrigens auch bei den zitierten Arbeiten von BURNETT und JENKINS der Fall.) Bei der Untersuchung

[1] Die Ursache der Verzögerung kann bei der Kiceschen Ableitung sowohl Übertragung als auch Kopolymerisation sein.

von Übertragungsreaktionen liegt das Hauptgewicht jedoch auf der Verminderung des Polymerisationsgrades. Es erschien uns daher angebracht, die von KICE gegebenen Ableitungen sinngemäß auf den Polymerisationsgrad zu übertragen.

2. Abgeänderte Kice-Gleichungen. Die Gleichung für die Polymerkette $\bar{\nu}'$ lautet bei Vorliegen von Verzögerung

$$\bar{\nu}' = \frac{v_w}{v_{ab} + v_{ab}' + \Sigma v_{ü}}. \qquad \text{(III,48)}$$

Die Reaktion III,46 (v_{ab}'') kommt hierin nicht vor, da durch diese keine Polymerketten abgebrochen werden.

Das Auftreten des zusätzlichen Abbruches III,45 (v_{ab}') bedingt nun aber, daß die allgemeine Übertragungsgleichung III,8 nicht mehr in der ursprünglichen Form gültig ist, sondern daß ein zusätzliches, subtraktives Glied auftritt.

Aus Gl. III,48 folgt nämlich, unter Einsetzen der entsprechenden Geschwindigkeitsgleichungen und Ersetzen von $[P^*]$ nach Gl. III,4 und Gl. III,6:

$$\frac{1}{\bar{\nu}'} - \frac{k_{ab}\, v_{Br}}{k_w^2\, [M]^2} - \frac{k_{ab}'[X^*]}{k_w[M]} = \Sigma \frac{v_{ü}}{v_w}. \qquad \text{(III,49)}$$

Man muß jetzt einen Ausdruck für die Konzentration der Überträgerradikale $[X^*]$ aufstellen. Man erhält ihn mit Hilfe der Radikalbilanz für den quasistationären Zustand, bei welcher nun alle radikalverbrauchenden Reaktionen, also auch v_{ab}'', berücksichtigt werden müssen. Wir gehen dabei in derselben Weise vor wie KICE (*70*), verwenden lediglich eine andere Definition für die Bruttogeschwindigkeit. (Bei KICE ist $v_{Br} \equiv -d \ln [M]/dt$; in der vorliegenden Arbeit $v_{Br} \equiv -d\,[M]/dt = v_w$ (vgl. Gl. I,11).

Im quasistationären Zustand ($d\,[P^*]/dt \cong 0$; $d\,[X^*]/dt \cong 0$) ist bei Verzögerung:

$$v_{st} = v_{ab} + v_{ab}' + v_{ab}''. \qquad \text{(III,50)}$$

Im Falle der unverzögerten Reaktion ist, wegen $v_{ab}' = v_{ab}'' = 0$:

$$v_{st} = v_{ab,0} = k_{ab}\,[P^*]_0^2. \qquad \text{(III,51)}$$

(Der Index 0 bezieht sich hier und im folgenden immer auf die unverzögerte Reaktion, d. h. auf die Polymerisation unter sonst gleichen Bedingungen, aber ohne Zusatz des Verzögerers.)

Aus III,50 und 51 folgt, unter Einsetzen der Geschwindigkeitsgleichungen und Ersetzen von $[P^*]$ nach Gl. III,4 und III,6:

$$\frac{k_{ab}\, v_{Br,0}^2}{k_w^2\, [M]^2} = \frac{k_{ab}\, v_{Br}^2}{k_w^2\, [M]^2} + \frac{k_{ab}'\, v_{Br}}{k_w[M]}\, [X^*] + k_{ab}''\, [X^*]^2. \qquad \text{(III,52)}$$

Aus dieser quadratischen Gleichung erhält man:

$$[X^*] = \frac{k_{ab}'\, v_{Br}}{2\, k_{ab}''\, k_w [M]} \left[\overset{+}{(-)} \sqrt{1 + c\left(\frac{v^2_{Br,0}}{v^2_{Br}} - 1\right)} - 1 \right]. \tag{III,53}$$

Wie bei KICE hat c dabei die Bedeutung:

$$c = \frac{k_{ab}\, k_{ab}''}{k_{ab}'^2}. \tag{III,54}$$

Durch Einsetzen von (III,53) in (III,49) erhalten wir schließlich:

$$\frac{1}{\bar{\nu}'} - \frac{k_{ab}\, v_{Br}}{k_w^2 [M]^2} (1 + \Omega) = \Sigma \frac{v_ü}{v_w}, \tag{III,55}$$

wobei

$$\Omega = \frac{2}{c} \left[\sqrt{1 + c\left(\frac{v^2_{Br,0}}{v^2_{Br}} - 1\right)} - 1 \right], \tag{III,56}$$

und

$$\Sigma \frac{v_ü}{v_w} = \left(\Sigma \frac{v_ü}{v_w}\right)_0 + C_x \frac{[X]}{[M]}. \tag{III,57}$$

Man kann nun analog vorgehen wie KICE, indem man für c Zahlenwerte einsetzt. Durch Ausprobieren muß c so gewählt werden, daß sich für eine gegebene Meß-Serie bei Auftragen von $\Sigma(v_ü/v_w)$ gegen $[X]/[M]$ nach Gl. III,57 die beste gerade Linie ergibt. (Über die Berechnung von $\bar{\nu}'$ aus dem jeweiligen experimentellen Mittelwert des Polymerisationsgrades siehe den nachfolgenden Abschnitt.)

Tabelle 2.
Hilfstabelle zur Berechnung von Übertragungskonstanten bei verzögerter Polymerisation
Ω als Funktion von $v_{Br,0}/v_{Br}$ und c nach Gl. III,56

$\frac{v_{Br,0}}{v_{Br}}$	$c = 1$	$c = 10^{-1}$	$c = 10^{-2}$	$c = 10^{-3}$	$c = 10^{-4}$	$c = 10^{-5}$
1,2	0,4	0,42	0,44	0,44	0,44	0,44
1,4	0,8	0,94	0,96	0,96	0,96	0,96
1,6	1,2	1,50	1,55	1,56	1,56	1,56
1,8	1,6	1,95	2,23	2,26	2,26	2,26
2,0	2,0	2,48	2,98	3,00	3,00	3,00
3,0	4,0	6,83	7,85	8,00	8,00	8,00
4,0	3,0	10,0	14,5	14,9	15,0	15,0
5,0	8,0	16,9	22,7	23,9	24,0	24,0
7,0	12,0	28,0	43,3	47,4	48,0	48,0
10,0	18,0	46,0	82,1	96,6	100,1	100,1

Die etwas langwierige Rechenarbeit des Ausprobierens verschiedener c-Werte kann durch Tab. 2 (oder eine entsprechende graphische Darstellung, die die Interpolation der Zwischenwerte erlaubt) erleichtert werden. Die obere Grenze von $c = 1$ ist wegen der oben bereits erwähnten

Kopolymerisationsversuche als sinnvoll anzusehen. Falls spezielle Versuche es erfordern, können jedoch auch $c > 1$ eingesetzt werden.

Wie man aus Tab. 2 ersehen kann, strebt Ω für geringe Verzögerungen mit kleiner werdendem c sehr schnell einem Grenzwert zu. Dies legt die Anwendung einer Näherungsformel für diese Fälle nahe.

Aus Gl. III,56 ist zu entnehmen, daß der zweite Summand unter der Wurzel für $(v_{Br,0}/v_{Br}) < 1{,}4$ und $c \lesseqgtr 1$ immer kleiner als 1 bleibt. Man kann die Wurzel daher in eine Reihe entwickeln und erhält als Näherung:

$$\Omega \cong \frac{v_{Br,0}^2}{v_{Br}^2} - 1 .$$

Wenn man dies in Gl. III,55 einsetzt, erhält man

$$\frac{1}{\bar{\nu}'} - \frac{k_{ab}\, v_{Br}}{k_w^2\,[M]^2}\,\frac{v_{Br,0}^2}{v_{Br}^2} = \Sigma\,\frac{v_{ü}}{v_w} . \qquad \text{(III,58)}$$

Gl. III,58 ist für Fälle, in denen die Verminderung der Geschwindigkeit nicht mehr als 30—40% beträgt, eine gute Näherung.

Aber auch für stärkere Verzögerungseffekte wird man sich häufig die langwierige Rechenarbeit des Ausprobierens von c-Werten in Gl. III,56 ersparen können, nämlich immer dann, wenn $1/\bar{\nu}' \gg k_{ab} v_{Br}/k_w^2\,[M]^2$. Man kann dann entweder das subtraktive Glied links in Gl. III,55 ganz vernachlässigen, oder aber durch Anwendung der beiden Grenzwerte von Ω für $c = 1$ und $c = 10^{-5}$* die Übertragungskonstante des Verzögerers in engen Grenzen festlegen.

Wenn das Lösungsmittel selbst der Verzögerer ist, bereitet die Definition der „unverzögerten Reaktion" Schwierigkeiten. Bekanntlich ist die Geschwindigkeit in Lösung von der Radikalausbeute (efficiency) des Intiators abhängig (*58, 16*) und kann sehr stark mit der Art des Lösungsmittels variieren. Matsumoto u. Mitarb. (*79*) haben bei der Polymerisation von Vinylacetat in verschiedenen Estern erhebliche Verzögerungen gefunden, konnten aber zeigen, daß diese auf eine besonders starke Verminderung der Radikalausbeute zurückzuführen sind. Man kann das leicht durch eine sinngemäße Anwendung der allgemeinen Übertragungsgleichung III,8 nachweisen, die wir hier nochmals aufschreiben:

$$\frac{1}{\bar{\nu}'} = \frac{k_{ab}\, v_{Br}}{k_w^2\,[M]^2} + C_M + C_{LM}\,\frac{[LM]}{[M]} .$$

Man trägt $1/\bar{\nu}'$ bei konstant gehaltenem Verhältnis $[LM]/[M]$ gegen $v_{Br}/[M]^2$ auf, wobei man die Geschwindigkeit durch Veränderung der Initiatorkonzentration variiert. $\bar{\nu}'$ wird dabei, je nach Abbruchsmechanismus, nach einer der in den vorangegangenen Abschnitten abgeleiteten Formeln durch den gemessenen Mittelwert des Polymerisationsgrades

* Bei $v_{Br,0}/v_{Br} > 10$ muß der untere Grenzwert erst errechnet werden.

ersetzt. (Bei Abbruch durch Kombination und Viscositätsmittel z. B. nach Gl. III,29.) Wenn die Verzögerung nur auf eine Verminderung der Startgeschwindigkeit zurückzuführen ist, müssen für verschiedene $[LM]/[M]$ parallele Geraden mit der Steigung k_{ab}/k_w^2 erhalten werden (siehe Abb. 25, S. 559). Aus den Ordinatenabschnitten kann man dann bei Kenntnis von C_M die Übertragungskonstante des Lösungsmittels bestimmen.

Wenn nicht parallele Kurven auftreten, ist die Verzögerung wahrscheinlich auf die Reaktionen III,45 und III,46 zurückzuführen. Man bezieht sich dann am besten auf die Polymerisation unter sonst gleichen Bedingungen in einem nicht verzögernden Lösungsmittel (z. B. Dioxan bei Styrol) und wertet nach III,58 aus. Wegen der bereits erwähnten möglichen Unterschiede in der Radikalausbeute werden die Ergebnisse dann jedoch nicht ganz eindeutig sein.

3. Das Ersetzen der Polymerkette $\bar{\nu}'$ durch den experimentellen Polymerisationsgrad bei Verzögerung. Zur Auswertung der Gl. III, 55 oder der Näherungsgleichung III,58 muß $\bar{\nu}'$ wieder durch den experimentell ermittelten Polymerisationsgrad ersetzt werden. Dies geschieht, wie in allen vorangegangenen Abschnitten, nach den Gleichungen III,9—11, wobei man wieder den Gewichtsanteil y der Moleküle mit *einer* Polymerkette für jeden Einzelfall berechnen muß.

Für *Disproportionierungs*abbruch ist in jedem Fall $y = 1$, und man erhält

$$\bar{\nu}' = \bar{P}_n \tag{III,59}$$

$$\bar{\nu}' = \bar{P}_w/2 \tag{III,60}$$

$$\bar{\nu}' = \bar{P}_\eta/(1 + b)\,. \tag{III,61}$$

Bei *Kombinations*abbruch sind in dem Bruchteil y die durch Übertragung (Reaktion III,43) sowie die durch die Radikale X^* (Reaktion III,45) abgebrochenen Moleküle enthalten. Nur die durch Kombination abgebrochenen Moleküle machen den Bruchteil $(1 - y)$ aus.

Es ist also:

$$y = \frac{\Sigma v_{ü} + v_{ab'}}{\Sigma v_{ü} + v_{ab'} + v_{ab}}$$

und

$$(1 - y) = \frac{v_{ab}}{\Sigma v_{ü} + v_{ab'} + v_{ab}}\,. \tag{III,62}$$

Wir verwenden nun wieder die bereits früher definierte Größe $\langle\nu\rangle$ (diejenige Kettenlänge, die auftreten würde, wenn nur der Kombinationsabbruch vorläge). Diese ist gegeben durch

$$\langle\bar{\nu}\rangle = \frac{v_w}{v_{ab}} = \frac{k_w^2\,[M]^2}{k_{ab}\,v_{Br}} \tag{III,63}$$

Formal ist $\langle\bar{\nu}\rangle$ durch dieselbe Gleichung gegeben, wie die kinetische Kettenlänge der unverzögerten Reaktion (vgl. III,7). Da jedoch bei Verzögerung $v_{Br} < v_{Br,0}$, ist $\langle\bar{\nu}\rangle$ größer als $\bar{\nu}_0$.

Aus (III,62), (III,48) und (III,63) folgt:

$$y = 1 - \frac{\bar{\nu}'}{\langle\bar{\nu}\rangle}.$$

Hierin ersetzen wir wieder $\bar{\nu}'$ durch die experimentellen Mittelwerte des Polymerisationsgrades nach (III,9—11) sowie $\langle\bar{\nu}\rangle$ durch denjenigen Polymerisationsgrad $\langle\bar{P}_0\rangle$, der entstehen würde, wenn nur der Kombinationsabbruch vorläge, nach (III,37—39). Man erhält dabei wieder die Gleichungen III,20—22. Das hat zur Folge, daß auch bei Vorliegen von Verzögerung das Ersetzen von $\bar{\nu}'$ durch den experimentellen Polymerisationsgrad, unter Berücksichtigung des Abbruchsmechanismus, in derselben Weise vorgenommen werden kann wie in den vorangegangenen Abschnitten. Der einzige Unterschied zu den Gleichungen, welche bei normaler Geschwindigkeit gültig sind, ist somit der Faktor $(1 + \Omega)$ bzw. $(v_{Br,0}/v_{Br})^2$ in dem subtraktiven Term der Übertragungsgleichung (vgl. Gl. III,55 und III,58 mit III,8).

f) Zusammenfassung und Hinweise zur Anwendung der abgeleiteten Übertragungsgleichungen

Zur Erleichterung der Anwendung der abgeleiteten Übertragungsgleichungen geben wir im folgenden eine Übersicht in Form von Tabellen, mit deren Hilfe man sofort die richtige Gleichung für jeden der behandelten Fälle zusammenstellen kann.

Um die Tabellierung zu ermöglichen, bezeichnen wir das subtraktive Glied in allen Übertragungsgleichungen (vgl. z. B. III,23 oder III,58) mit B. Wir schreiben die allgemeine Übertragungsgleichung (vgl. III,8) also in der Form:

$$\frac{A}{\bar{P}} - B = \Sigma \frac{v_{ü}}{v_w}.$$

Da zur Berechnung der A die Rechengrößen $\langle\bar{P}_0\rangle$ benötigt werden, sind auch diese in die Tabellen mit aufgenommen.

Die Aufteilung erfolgt nach den drei verschiedenen Mittelwerten des Polymerisationsgrades.

Die Gleichungen für die unverzögerte Polymerisation (a, b und c in den Tabellen) können immer dann angewendet werden, wenn die Geschwindigkeit durch die Übertragungsreaktion nicht beeinflußt wird, oder aber, wenn durch geeignete Versuche festgestellt werden kann, daß eine beobachtete Verminderung der Bruttogeschwindigkeit ausschließlich durch Verringerung der Startgeschwindigkeit hervorgerufen wurde (vgl. Abschnitt III, e).

Für die *verzögerte* Polymerisation wurden in den Tabellen die Näherungsgleichungen verwendet, welche jedoch nur bis zu Verminderungen

Tabelle 3. *Übertragungsgleichungen für das Zahlenmittel des Polymerisationsgrades*

$$\frac{A_n}{\overline{P}_n} - B_n = \Sigma \frac{v_{ü}}{v_w}$$

		A_n	$\langle \overline{P}_{n,0} \rangle$	B_n
a	Disproportionierung	1	—	$\frac{k_{ab}\, v_{Br}}{k_w^2\,[M]^2}$
b	Kombination	$1 + \frac{\overline{P}_n}{\langle \overline{P}_{n,0} \rangle}$	$\frac{2\,k_w^2\,[M]^2}{k_{ab}\, v_{Br}}$	$\frac{k_{ab}\, v_{Br}}{k_w^2\,[M]^2}$
c	Disproportionierung und Kombination	$1 + \frac{\overline{P}_n}{\langle \overline{P}_{n,0} \rangle}$	$\frac{2\,k_w^2\,[M]^2}{k_{ab}\, v_{Br}}\,(\delta + 1)$	$\frac{k_{ab}\, v_{Br}}{k_w^2\,[M]^2}$
d	Disproportionierung Verzögerung	1	—	$\frac{k_{ab}\, v_{Br}}{k_w^2\,[M]^2} \cdot \frac{v_{Br,0}^2}{v_{Br}^2}$
e	Kombination Verzögerung	$1 + \frac{\overline{P}_n}{\langle \overline{P}_{n,0} \rangle}$	$\frac{2\,k_w^2\,[M]^2}{k_{ab}\, v_{Br}}$	$\frac{k_{ab}\, v_{Br}}{k_w^2\,[M]^2} \cdot \frac{v_{Br,0}^2}{v_{Br}^2}$
f	Disproportionierung und Kombination Verzögerung	$1 + \frac{\overline{P}_n}{\langle \overline{P}_{n,0} \rangle}$	$\frac{2\,k_w^2\,[M]^2}{k_{ab}\, v_{Br}}\,(\delta + 1)$	$\frac{k_{ab}\, v_{Br}}{k_w^2\,[M]^2} \cdot \frac{v_{Br,0}^2}{v_{Br}^2}$

Tabelle 4. *Übertragungsgleichungen für das Gewichtsmittel des Polymerisationsgrades*

$$\frac{A_w}{\overline{P}_w} - B_w = \Sigma \frac{v_{ü}}{v_w}$$

		A_w*	$\langle \overline{P}_{w,0} \rangle$	B_w
a	Disproportionierung	2	—	$\frac{k_{ab}\, v_{Br}}{k_w^2\,[M]^2}$
b	Kombination	$1 + \sqrt{1 + 3\,\overline{P}_w/\langle \overline{P}_{w,0} \rangle}$	$\frac{3\,k_w^2\,[M]^2}{k_{ab}\, v_{Br}}$	$\frac{k_{ab}\, v_{Br}}{k_w^2\,[M]^2}$
c	Disproportionierung und Kombination	$1 + \sqrt{1 + 3\,\overline{P}_w/\langle \overline{P}_{w,0} \rangle}$	$\frac{3\,k_w^2\,[M]^2}{k_{ab}\, v_{Br}}\,(\delta + 1)$	$\frac{k_{ab}\, v_{Br}}{k_w^2\,[M]^2}$
d	Disproportionierung Verzögerung	2	—	$\frac{k_{ab}\, v_{Br}}{k_w^2\,[M]^2} \cdot \frac{v_{Br,0}^2}{v_{Br}^2}$
e	Kombination Verzögerung	$1 + \sqrt{1 + 3\,\overline{P}_w/\langle \overline{P}_{w,0} \rangle}$	$\frac{3\,k_w^2\,[M]^2}{k_{ab}\, v_{Br}}$	$\frac{k_{ab}\, v_{Br}}{k_w^2\,[M]^2} \cdot \frac{v_{Br,0}^2}{v_{Br}^2}$
f	Disproportionierung und Kombination Verzögerung	$1 + \sqrt{1 + 3\,\overline{P}_w/\langle \overline{P}_{w,0} \rangle}$	$\frac{3\,k_w^2\,[M]^2}{k_{ab}\, v_{Br}}\,(\delta + 1)$	$\frac{k_{ab}\, v_{Br}}{k_w^2\,[M]^2} \cdot \frac{v_{Br,0}^2}{v_{Br}^2}$

* In den Fällen b, c, e und f kann A_w auch aus Abb. 1 entnommen werden.

Tabelle 5. *Übertragungsgleichungen für das Viskositätsmittel des Polymerisationsgrades* (a = Exponent der Viscositäts-Molekulargewichtsbeziehung, Gl. II,17)

$$\frac{A_\eta}{\overline{P}_\eta} - B_\eta = \Sigma \frac{v_{ü}}{v_w}$$

		A_η	$\langle \overline{P}_{\eta,0} \rangle$	B_η
a	Disproportionierung	$\frac{3+a}{2}$	—	$\frac{k_{ab}\, v_{Br}}{k_w^2\,[M]^2}$
b	Kombination	aus Abb. 2	$\frac{5+a}{2} \cdot \frac{k_w^2\,[M]^2}{k_{ab}\, v_{Br}}$	$\frac{k_{ab}\, v_{Br}}{k_w^2\,[M]^2}$
c	Disproportionierung und Kombination	aus Abb. 2	$\frac{5+a}{2} \cdot \frac{k_w^2\,[M]^2}{k_{ab}\, v_{Br}}\,(\delta+1)$	$\frac{k_{ab}\, v_{Br}}{k_w^2\,[M]^2}$
d	Disproportionierung Verzögerung	$\frac{3+a}{2}$	—	$\frac{k_{ab}\, v_{Br}}{k_w^2\,[M]^2} \cdot \frac{v_{Br,0}^2}{v_{Br}^2}$
e	Kombination Verzögerung	aus Abb. 2	$\frac{5+a}{2} \cdot \frac{k_w^2\,[M]^2}{k_{ab}\, v_{Br}}$	$\frac{k_{ab}\, v_{Br}}{k_w^2\,[M]^2} \cdot \frac{v_{Br,0}^2}{v_{Br}^2}$
f	Disproportionierung und Kombination Verzögerung	aus Abb. 2	$\frac{5+a}{2} \cdot \frac{k_w^2\,[M]^2}{k_{ab}\, v_{Br}}\,(\delta+1)$	$\frac{k_{ab}\, v_{Br}}{k_w^2\,[M]^2} \cdot \frac{v_{Br,0}^2}{v_{Br}^2}$

der Geschwindigkeit von etwa 40% gebraucht werden sollten. Bei stärkerer Verzögerung sei auf den Abschnitt III, e verwiesen.

Über die Bestimmung von $\delta \equiv k_{dis}/k_{com}$ siehe Abschnitt IV, a.

Wie man aus den Tabellen sehen kann, sind die Gleichungen für reinen Disproportionierungsabbruch besonders einfach. Zwar ist bisher kein Polymeres bekannt, von dem mit Sicherheit angenommen werden könnte, daß die einzige Kettenabbruchsreaktion Disproportionierung sei. Trotzdem können die hierfür abgeleiteten Gleichungen, deren einzige Voraussetzung $y = 1$ ist, immer dann angewendet werden, wenn die Übertragungseffekte so stark sind, daß man etwa vorhandenen Kombinationsabbruch daneben vernachlässigen kann ($y \to 1$). Dies ist z. B. bei Vinylacetat fast immer der Fall.

Die in den Tabellen 2—4 zusammengefaßten Gleichungen liefern in jedem Fall $\Sigma(v_{ü}/v_w)$ aus *Einzelversuchen*.

Die Summe aller Übertragungsterme besteht aus ebensovielen Summanden, wie Überträgersubstanzen im System vorhanden sind:

$$\Sigma \frac{v_{ü}}{v_w} = C_M + C_{LM} \frac{[LM]}{[M]} + C_I \frac{[I]}{[M]} + C_X \frac{[X]}{[M]} + \cdots \qquad \text{(III,64)}$$

Zur Bestimmung der *Übertragungskonstanten* führt man am besten Meßreihen durch, bei welchen man die Konzentration des zu untersuchenden Überträgers variiert und dafür sorgt, daß die übrigen Summanden in Gl. III,64 konstant gehalten werden.

IV. Zur Bestimmung einiger Hilfsgrößen

a) Das Verhältnis von Disproportionierungs- zu Kombinationsabbruch

Von MELVILLE u. Mitarb. (*19*) sowie von ALLEN u. Mitarb. (*1*) wurde nachgewiesen, daß bei *Polymethylmethacrylat* die beiden Abbruchmechanismen, Kombination und Disproportionierung, nebeneinander vorliegen. Es gibt unseres Wissens in der Literatur noch keine Angaben darüber, ob auch bei anderen Monomeren beide Abbrüche vorkommen. Bekannt ist jedoch, daß bei *Styrol* der Abbruch nur durch Kombination erfolgt (*19, 60*).

Das Verhältnis von Disproportionierung zu Kombination wird dadurch ermittelt, daß man die Polymerisation mit radioaktiv markiertem Initiator startet und dann die durchschnittliche Anzahl der radioaktiven Endgruppen pro Polymermolekül bestimmt. Disproportionierungsmoleküle enthalten eine, Kombinationsmoleküle zwei Endgruppen. In die Ausrechnung der Endgruppenzahl geht das Zahlenmittel des Polymerisationsgrades, $\overline{P}_n$, ein, welches durch Messung des osmotischen Druckes bestimmt wird. Die erwähnten Autoren (*19, 1*) haben diese Messungen am unfraktionierten Polymerisat vorgenommen. Hierbei muß man jedoch eine erhebliche Unsicherheit in Kauf nehmen, da bei unbekannter Durchlässigkeit der Membran für die niederen Polymeren ein unbestimmter, von $\overline{P}_n$ abweichender Mittelwert erhalten werden kann (*120, 91, 132*). Die Ergebnisse der beiden zitierten Arbeiten zeigen auch tatsächlich erhebliche Differenzen.

Ein Verfahren, über welches kürzlich berichtet wurde (*131*) und bei welchem die Molekulargewichtsbestimmungen an Fraktionen durchgeführt werden, dürfte prinzipiell zuverlässigere Werte für das Verhältnis der beiden Abbruchreaktionen liefern.

Unter Rückgriff auf bereits in der vorliegenden Arbeit verwendete Gleichungen soll diese Methode kurz beschrieben und das Ergebnis an Methacrylat wiedergegeben werden.

Bei Methylmethacrylat sind die Übertragungskonstanten des Monomeren, des Polymeren sowie der üblichen Lösungsmittel sehr niedrig (siehe Abschnitt V). Man kann sie daher mit guter Näherung vernachlässigen, solange der Polymerisationsgrad verhältnismäßig niedrig gehalten wird ($1/\overline{P} \gg \Sigma v_{ü}/v_w$). Die Polymerkette $\bar{\nu}'$ kann dann der kinetischen Kettenlänge $\bar{\nu}$ gleichgesetzt werden.

Unter dieser Voraussetzung ist der Bruchteil y derjenigen Moleküle, welche aus einer Kette bestehen, gegeben durch die Gleichung

$$y = \frac{v_{dis}}{v_{dis} + v_{com}}. \qquad \text{(IV,1)}$$

Da der Kettenabbruch durch Disproportionierung bzw. durch Kombination jeweils eine bimolekulare Reaktion zweier wachsender Ketten ist, kann man in Gl. IV,1 anstelle der Geschwindigkeiten die Geschwindigkeitskonstanten setzen:

$$y = \frac{k_{dis}}{k_{dis} + k_{com}}. \tag{IV,1a}$$

Die aus einer Kette bestehenden Disproportionierungsmoleküle haben nun eine andere Molekulargewichtsverteilung als die aus 2 Ketten bestehenden kombinierten Moleküle; die beiden Verteilungen ergeben sich durch Einsetzen von $k = 1$ bzw. $k = 2$ in Gl. II,6. In Abb. 3 sind die beiden Verteilungen dargestellt, wobei, als Beispiel, $y = 0{,}5$, d. h. $k_{dis} = k_{com}$ zugrunde gelegt wurde. Um die Kurven allgemeingültig zu halten, wurde als Abszisse der mit der kinetischen Kettenlänge reduzierte Polymerisationsgrad, $P/\bar{\nu}$, verwendet. Die Ordinate ist daher nicht die Massenverteilung m_P nach Gl. II,6, sondern eine entsprechend veränderte Größe m_P^*, welche die Summe der beiden Flächen auf den Wert eins normiert.

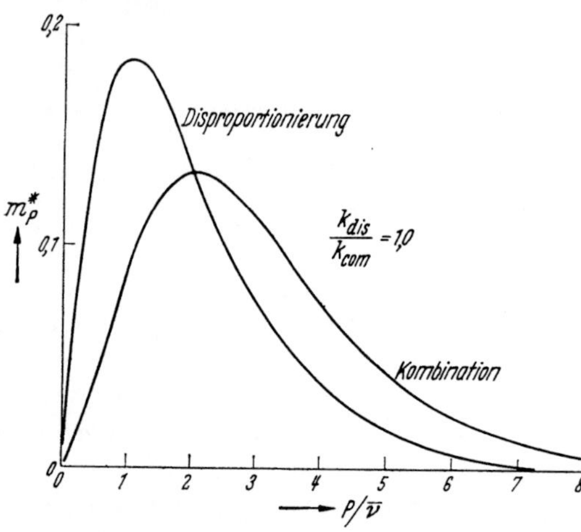

Abb. 3. Relative Molekulargewichtsverteilung der Disproportionierungs- bzw. der Kombinationsmoleküle für $y = 0{,}5$ (*131*)

Aus den Überschneidungen der beiden Kurven und aus der Verschiedenheit der Endgruppenzahlen Z für die beiden Molekülsorten (Kombination $Z = 2$, Disproportionierung $Z = 1$) ergibt sich, daß für eine Reihe von Fraktionen desselben Polymerisates die mittlere Endgruppenzahl mit zunehmendem Molekulargewicht der Fraktionen ansteigt.

Wir bezeichnen den Massenanteil jener Moleküle vom Polymerisationsgrad P, die durch Disproportionierung abgebrochen wurden, mit $m_{P,1}$, hingegen den Massenanteil der Moleküle vom selben Polymerisationsgrad P, welche durch Kombination entstanden sind, mit $m_{P,2}$. Die mittlere Zahl der Endgruppen $Z(P)$ sämtlicher Moleküle vom Polymerisationsgrad P ist dann:

$$Z(P) = \frac{y\, m_{P,1} + 2(1-y)\, m_{P,2}}{y\, m_{P,1} + (1-y)\, m_{P,2}}.$$

Setzen wir hierin die beiden Massenanteile nach Gl. II,6 ein ($m_{P,1}$: $k = 1$; $m_{P,2}$: $k = 2$), so erhalten wir:

$$Z(P) = \frac{y + (1-y)(-\ln\alpha)\, P}{y + (1-y)(-\ln\alpha)\, P/2}.$$

Nach Gl. II,2 und 3 ist $\alpha = 1 - v_{ab}/v_w$. Da $v_{ab} \ll v_w$, gilt

$$-\ln \alpha = v_{ab}/v_w = 1/\bar{\nu}\,. \tag{IV,2}$$

Damit erhalten wir:

$$Z(P) = \frac{y + (1-y)\,P/\bar{\nu}}{y + (1-y)\,P/2\bar{\nu}}\,.$$

Anstelle von $\bar{\nu}(=\bar{\nu}')$ kann nun der experimentelle Polymerisationsgrad des Gesamtpolymeren eingesetzt werden unter Verwendung der früher abgeleiteten Gleichungen III,9—11. Bei Verwendung des *Zahlenmittels* erhält man:

$$Z(P) = \frac{1 + 2w_n P/\bar{P}_n}{1 + w_n P/\bar{P}_n} \tag{IV,3}$$

mit

$$w_n = \frac{1-y}{y(1+y)}\,. \tag{IV,3a}$$

Für das *Gewichtsmittel* ergibt sich:

$$Z(P) = \frac{1 + 2w_w P/\bar{P}_w}{1 + w_w P/\bar{P}_w} \tag{IV,4}$$

mit

$$w_w = \frac{3 - 4y + y^2}{2y} \tag{IV,4a}$$

und schließlich für das *Viscositätsmittel*

$$Z(P) = \frac{1 + 2w_\eta P/\bar{P}_\eta}{1 + w_\eta P/\bar{P}_\eta} \tag{IV,5}$$

mit

$$w_\eta = \frac{1-y}{2y(y+1)}\,(2 + b + 2by - by^2)\,. \tag{IV,5a}$$

Aus den bereits erwähnten Gründen empfiehlt es sich nicht, das Zahlenmittel des Polymerisationsgrades des gesamten Polymeren zu messen. Meist wird als zuverlässigster Durchschnitt das Viscositätsmittel zur Verfügung stehen.

In Abb. 4 ist $Z(P)$ als Funktion von $P/\bar{P}_\eta$ für verschiedene Werte von y dargestellt. In diese, theoretisch nach Gl. IV,5 berechnete Kurvenschar sind Meßwerte eingezeichnet, die an Fraktionen von Polymethylmethacrylat-Präparaten gewonnen wurden, welche bei verschiedener Temperatur hergestellt sind.

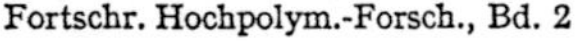

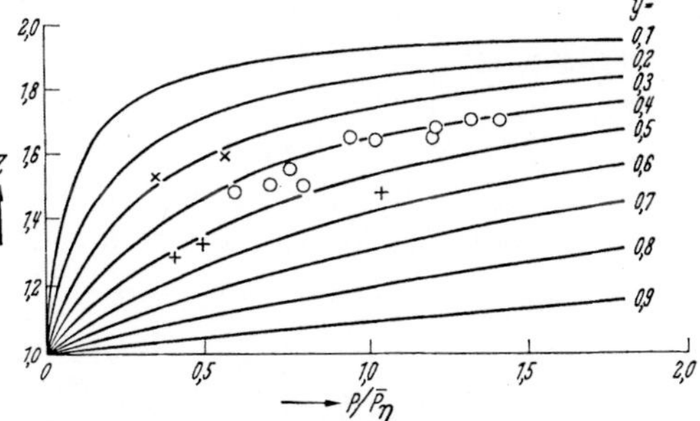

Abb. 4. Mittlere Endgruppenzahl in Abhängigkeit von $P/\bar{P}_\eta$ für verschiedene Werte von y nach Gl. IV,5, wobei $b = 0{,}88$. Eingezeichnete Meßpunkte von Fraktionen von Polymethylmethacrylaten (*131*). × : 40° C; ○ : 60° C; + : 80° C

Aus der Lage der Meßpunkte kann y für die verschiedenen Temperaturen direkt abgeschätzt werden. [Genauer ist eine rechnerische Auswertung von Gl. IV,5 (*131*).]

Das Verhältnis der beiden Abbruchskonstanten läßt sich nun, unter Berücksichtigung von Gl. IV,1, ausrechnen nach der Gleichung:

$$\delta \equiv \frac{k_{dis}}{k_{com}} = \frac{y}{1-y} . \quad \text{(IV,6)}$$

Aus den in Abb. 4 dargestellten Meßwerten an Polymethacrylaten wurde die Temperaturabhängigkeit des Verhältnisses k_{dis}/k_{com} für dieses Polymere bestimmt als:

$$\log_{10} \delta \equiv \log_{10} (k_{dis}/k_{com}) = 2.89 - 1000/T . \quad \text{(IV,7)}$$

b) Die Geschwindigkeitskonstanten von Wachstum und Abbruch (k_w^2/k_{ab})

Zur Bestimmung des Konstantenverhältnisses k_w^2/k_{ab} gibt es prinzipiell zwei Wege. Der eine beruht auf der Messung von Bruttogeschwindigkeit und Startgeschwindigkeit, der zweite auf der Messung von Bruttogeschwindigkeit und Polymerisationsgrad.

Eine Voraussetzung für beide Methoden ist, daß die Startgeschwindigkeit gleich der Abbruchsgeschwindigkeit ist (Stationaritätsbedingung):

$$v_{st} = v_{ab} . \quad \text{(IV,8)}$$

Daraus folgt, unter Berücksichtigung von (I,15):

$$[P^*] = v_{st}^{1/2} k_{ab}^{-1/2} .$$

Setzt man dies in die Gleichung für die Bruttogeschwindigkeit (Gl. I,11) ein, so erhält man

$$v_{Br} = \frac{k_w}{k_{ab}^{1/2}} v_{st}^{1/2} [M] , \quad \text{(IV,9)}$$

und somit:

$$\frac{k_w^2}{k_{ab}} = \frac{v_{Br}^2}{v_{st} [M]^2} . \quad \text{(IV,10)}$$

Die Bestimmung der Startgeschwindigkeit geschieht am besten nach einer zuerst von BEVINGTON (*16*) verwendeten Methode, wonach man, unter Benützung von C^{14}-markiertem Initiator, die Zahl der Initiatorfragmente bestimmt, die in einer bestimmten Menge Polymerem enthalten sind. Bei Kenntnis der Polymerisationsgeschwindigkeit läßt sich dann leicht v_{st} berechnen. Diese Methode hat den Vorteil, daß der Polymerisationsgrad nicht benötigt wird. Sie hat jedoch folgende Unsicherheiten:

1. Wenn die Initiatorradikale nicht nur starten, sondern auch Ketten abbrechen, wie dies z. B. bei Azobisisobutyronitril (AIBN) in Styrol der Fall ist (*7a*, *58*), werden zu hohe Werte für v_{st} gefunden.

2. In manchen Fällungsmitteln (z. B. Benzin) ist die Löslichkeit von AIBN relativ schlecht; es ist dann sehr schwer, das Präparat restlos von nicht eingebautem Initiator zu befreien.

3. Der Verlust niedermolekularer Anteile beim Aus- und Umfällen kann zu Fehlern führen. [Polymethacrylat vom Molekulargewicht 5000 läßt sich z. B. aus Benzol mit Methanol nicht ausfällen (*61*).]

Der Vollständigkeit halber sei auch die Bestimmung der Startgeschwindigkeit mit Hilfe der Inhibierung durch stabile freie Radikale erwähnt, wie sie z. B. von Bartlett und Kwart (*9*) oder von Matheson u. Mitarb. (*76*) durchgeführt wurde. Tüdös u. Mitarb. (*146*) haben jedoch gezeigt, daß die üblicherweise hierfür verwendeten Radikale Nebenreaktionen mit dem Monomeren eingehen. Diese Autoren weisen darauf hin, daß nur unter sorgfältiger Berücksichtigung dieser Nebenreaktionen die Startgeschwindigkeit exakt bestimmt werden kann.

Die *zweite* Methode verwertet den Polymerisationsgrad. Die Frage der Mittelwertsbildung der verwendeten Bestimmungsmethode muß daher hier wieder berücksichtigt werden. Die obenerwähnten Unsicherheiten der ersten Methode können jedoch weitgehend vermieden werden. Wie kürzlich am Beispiel von Styrol und AIBN gezeigt wurde, kann k_w^2/k_{ab} unter quantitativer Berücksichtigung des Kettenabbruches durch Primärradikale bestimmt werden (*58*). Bei Verwendung des Viscositätsmittelwertes des Polymerisationsgrades wird ferner etwaiger Verlust niedermolekularer Anteile beim Ausfällen weniger bedeutungsvoll, da diese Art von Durchschnittswert (wie auch das Gewichtsmittel) sehr viel weniger empfindlich gegenüber den niedermolekularen Anteilen ist als etwa das Zahlenmittel oder die obenerwähnte Zählung der eingebauten Initiatorreste. In extremen Fällen sehr niederer Polymerisationsgrade wird man jedoch vorteilhaft die Viscositätszahl direkt in der Reaktionslösung, ohne vorheriges Ausfällen, bestimmen (*129*) (siehe nächsten Abschnitt).

Sofern die Initiatorradikale nicht kettenabbrechend wirken und Übertragung vernachlässigt werden kann ($\bar{\nu} = \bar{\nu}'$), folgt aus (III,7):

$$\frac{k_w^2}{k_{ab}} = \frac{v_{Br} \cdot \bar{\nu}'}{[M]^2}.$$

Bei Vorliegen von Abbruch durch Primärradikale wird daraus (*58*):

$$\frac{k_w^2}{k_{ab}} = \frac{v_{Br} \cdot \bar{\nu}'}{[M]^2} \cdot \frac{1}{1 - g/f}, \qquad \text{(IV,11)}$$

wobei f und g die Bruchteile der Initiatorradikale sind, die Ketten starten bzw. abbrechen.

In Gl. IV,11 muß nun wieder $\bar{\nu}'$ durch den gemessenen Polymerisationsgrad ersetzt werden. Verwendet man das Viscositätsmittel (Gl. II,23), so ist:

$$\frac{k_w^2}{k_{ab}} = \frac{v_{Br}\,\overline{P}_\eta}{[M]^2} \cdot \frac{(y+1)}{2 + b + 2by - by^2} \cdot \frac{1}{1 - g/f}.$$

Zur Bestimmung des Gewichtsbruches y muß berücksichtigt werden, welche Reaktionen zu Molekülen mit einer Polymerkette führen können.

Bei Styrol z. B. erfolgt der normale Kettenabbruch stets durch Kombination (*19, 60*). Bei Vermeidung von Übertragungseinflüssen ($1/P \gg \Sigma v_{ü}/v_w$) ist daher der Abbruch durch Primärradikale die einzige Reaktion, die berücksichtigt werden muß. [Es ist dann $y = g/f$ (*58*).]

Bei Methylmethacrylat konnte nachgewiesen werden (*131*), daß der Initiator AIBN keinen merkbaren Abbruch durch Initiatorradikale verursacht. y ist daher nur durch den Disproportionierungsanteil bestimmt.

Bei Vinylacetat ist bisher nichts darüber bekannt, ob bei niederen Temperaturen auch z. T. Kombinationsabbruch auftritt. Wegen der relativ hohen Werte aller Übertragungskonstanten (C_M ist z. B. bei 60° C bereits von der Größenordnung 10^{-4}, siehe Abschnitt V) dürfte der weitaus größte Teil aller Moleküle aus einer Kette bestehen. Man kann daher bei diesem Monomeren sicher keinen großen Fehler machen, wenn man grundsätzlich mit $y = 1$ rechnet.

Im einzelnen sei auf die Originalarbeiten verwiesen. Wir zitieren hier lediglich folgende Ergebnisse:

Für Styrol (*58*) wurde gefunden

$$\log_{10} k_w^2/k_{ab} = 4.26 - 2475/T\,, \tag{IV,12}$$

für Methylmethacrylat (*131*)

$$\log_{10} k_w^2/k_{ab} = 3.45 - 1810/T\,, \tag{IV,13}$$

für Vinylacetat (*4*)

$$\log_{10} k_w^2/k_{ab} = 14.3 - 4910/T\,. \tag{IV,14}$$

Die Gl. IV,14 für Vinylacetat wurde aus Meßergebnissen von AUTRATA und MÜLLER (*4*) neu berechnet. Die Autoren verwenden zur Bestimmung der Polymerisationsgrade eine Eichgleichung für die Viscositätszahl, welche von WAGNER (*149*) mit Hilfe von Fraktionen aufgestellt wurde. Der von den Autoren als Zahlenmittel angegebene Polymerisationsgrad ist daher in Wirklichkeit das Viscositätsmittel (vgl. Abschnitte Ia und Vc). Ferner wird die ungebräuchliche Konzentrationsangabe Mol/kg verwendet. Unter Umrechnung der letzteren auf Mol/l sowie unter Berücksichtigung von $\overline{P}_\eta/\overline{P}_n = 1{,}85$ (vgl. Abschnitt Vc) wurde Gl. IV,14 erhalten.

MATSUMOTO u. Mitarb. (*79*) geben für 60° C den Wert $k_w^2/k_{ab} = 0{,}385$, welcher sehr gut durch Gl. IV,14 wiedergegeben wird.

Mit Hilfe der von MATHESON u. Mitarb. (*74, 76, 75*) mit der Methode des rotierenden Sektors bestimmten Werte von k_w/k_{ab} und ihrer Temperaturabhängigkeiten lassen sich die Konstanten von Wachstum und Abbruch einzeln berechnen. Man erhält:

für Styrol (*58*)

$$\log_{10} k_w = 6{,}34 - 1295/T \tag{IV,12a}$$

$$\log_{10} k_{ab} = 8{,}42 - 115/T \tag{IV,12b}$$

für Methylmethacrylat (*131*)

$$\log_{10} k_w = 5{,}90 - 1060/T \qquad \text{(IV,13a)}$$

$$\log_{10} k_{ab} = 8{,}35 - 310/T \qquad \text{(IV,13b)}$$

für Vinylacetat

$$k_w\ (60^\circ\ \mathrm{C}) = 1{,}5 \cdot 10^4 \qquad \text{(IV,14a)}$$

$$k_{ab}\ (60^\circ\ \mathrm{C}) = 6{,}0 \cdot 10^8 \qquad \text{(IV,14b)}$$

c) Zur Bestimmung von $\overline{P}_\eta$

Die Viscositätszahl $[\eta]$ eines hochmolekularen Stoffes ist definiert als der Grenzwert der reduzierten spezifischen Viscosität für die Konzentration und den Gradienten Null.

$$[\eta] = \left[\frac{\eta_{sp}}{c}\right]_{\substack{c=0\\G=0}} \qquad \text{(IV,15)}$$

Dabei sind:

$\eta_{sp} \equiv (\eta_{LÖS} - \eta_{LM})/\eta_{LM}$ = spezifische Viscosität des Polymeren in Lösung

c = Konzentration des Polymeren $[g \cdot \mathrm{cm}^3]$

G = Geschwindigkeitsgradient $[\mathrm{sec}^{-1}]$

Theoretisch ist die Extrapolation auf den Gradienten Null notwendig, da statistische Fadenmoleküle mit wachsendem Geschwindigkeitsgradienten die Tendenz zeigen, sich abhängig von ihrer „inneren“ Viscosität (*72*) zu deformieren und bevorzugt in Strömungsrichtung zu orientieren. Der Grenzwert für η_{sp}/c für $c = 0$ liegt daher bei einem endlichen Gradienten tiefer als die durch Gl. IV,15 definierte Viscositätszahl. Praktisch wird jedoch in den meisten Fällen bei einem Gradienten von 1000 sec^{-1} bereits der Grenzwert für $G = 0$ erreicht (*114*)[1]. Da bei Capillarviscosimetern, unter Einhaltung gewisser Konstruktionsvorschriften, ein Gradient dieser Größe leicht zu erhalten ist, ist es deshalb üblich, die Viscositätszahl in solchen Viscosimetern zu bestimmen.

Die Extrapolation zur Konzentration Null kann entweder graphisch aus einer Konzentrationsreihe erfolgen oder rechnerisch aus einer einzigen Messung nach der Gleichung

$$[\eta] = \frac{\eta_{sp}/c}{1 + K\eta_{sp}}. \qquad \text{(IV,16)}$$

[1] Bei sehr hohen Polymerisationsgraden ist dies jedoch nicht der Fall (*135*, *33*). Wie G. V. SCHULZ und H. J. CANTOW (*124*) zeigten, kommt man jedoch trotzdem zu einer zuverlässigen Molekulargewichtsbestimmung, wenn man eine Normierung der Viscosimeter vornimmt. Man erhält dann eine experimentell festgelegte, von jeder hypothetischen Annahme unabhängige Beziehung zwischen Molekulargewicht und Viscosität.

Diese Gleichung wurde empirisch von Schulz und Blaschke (*122*) gefunden. Huggins (*65*) machte später den Versuch, sie theoretisch zu begründen.

Die Konstante K in (IV,16) wurde von verschiedenen Autoren für eine Reihe von Polymer-Lösungsmittel-Systemen bestimmt. Aus den Daten der Tab. 6 ist zu entnehmen (besonders bei Berücksichtigung der Streuung für dasselbe System), daß K in guten Lösungsmitteln[1] weitgehend vom speziellen System unabhängig zu sein scheint und daß der Mittelwert $K = 0{,}3$, zumindest in den vorliegenden Fällen, allgemein anwendbar sein dürfte. Voraussetzung für die Gültigkeit von Gl. IV,16 ist jedoch, daß $0{,}1 < \eta_{sp} < 0{,}4$.

Tabelle 6. *Die Konstante K aus Gl. IV,16 für verschiedene Polymer-Lösungsmittelsysteme*

Polymeres	Lösungsmittel	K (Gl. IV, 16)	Zitat
Polymethylmethacrylat	Chloroform	0,29	*134*
Polymethylmethacrylat	Benzol	0,25	*110*
Polymethylmethacrylat	Benzol	0,34	*142*
Polystyrol	Chloroform	0,36	*11*
Polystyrol	Benzol	0,29	*142*
Polystyrol	Toluol	0,27	*134*
Polystyrol	Toluol	0,38	*11*
Polystyrol	Butanon	0,29	*134*
Polyvinylacetat	Benzol	0,25	*142*
Polyvinylacetat	Butanon	0,31	*138*

Das Polymere, dessen Viscositätszahl bestimmt werden soll, wird im allgemeinen ausgefällt, getrocknet und mit bekannter Konzentration wieder gelöst. Falls bei einem Experiment extrem niedere Polymerisationsgrade zu erwarten sind, empfiehlt es sich, die Viscositätszahl direkt in der Reaktionslösung, ohne vorheriges Ausfällen, zu messen, um Verluste der niedermolekularen Anteile zu verhindern. Man muß dazu in Dilatometern polymerisieren, um mit Hilfe des Umsatzes die Konzentration genau bestimmen zu können (*129*).

Zur *Eichung* der Viscositätszahl verwendet man am besten die Geschwindigkeits-Ultrazentrifuge in Verbindung mit der Diffusion, da diese Methode praktisch in gleicher Weise mittelt wie die Viscosität (*88*).

Theoretisch sollte man, sofern man die Eichung mit Fraktionen durchführt, jede Methode der Molekulargewichtsbestimmung verwenden können, da für eine ideale Fraktion $P_w = P_\eta = P_n$. Praktisch ist bei Fraktionen jedoch kaum eine bessere Uneinheitlichkeit als 0,1—0,2 zu erreichen. Die Ultrazentrifugen-Methode ist daher unbedingt vorzuziehen. Sie hat zudem den Vorteil großer experimenteller Sicherheit.

[1] In schlechten Lösungsmitteln werden höhere Werte für K gefunden.

Für *Polystyrol* wurde von G. MEYERHOFF (*90*) folgende Beziehung zwischen der Viscositätszahl (bei 20° in Benzol) und dem Polymerisationsgrad aufgestellt:

$$[\eta] = 0{,}348\ \bar{P}_\eta^{0{,}72} \quad [\mathrm{cm^3\,g^{-1}}]\,. \qquad \text{(IV,17)}$$

Bei *Polymethylmethacrylat* fanden G. V. SCHULZ und G. MEYERHOFF (*133*) ($[\eta]$ bei 20° in Benzol):

$$[\eta] = 0{,}192\ \bar{P}_\eta^{0{,}76} \quad [\mathrm{cm^3\,g^{-1}}]\,. \qquad \text{(IV,18)}$$

Von H. G. ELIAS und F. PATAT (*43*) wurde die entsprechende Beziehung für *Polyvinylacetat* angegeben ($[\eta]$ bei 25° in Butanon):

$$[\eta] = 0{,}667\ \bar{P}_\eta^{0{,}62} \quad [\mathrm{cm^3\,g^{-1}}]\,. \qquad \text{(IV,19)}$$

Eine weitere Eichgleichung für *Polyvinylacetat* wurde von STOCKMAYER u. Mitarb. (*104*) gegeben ($[\eta]$ bei 25° in Aceton):

$$[\eta] = 0{,}312\ \bar{P}_\eta^{0{,}71} \quad [\mathrm{cm^3\,g^{-1}}]\,. \qquad \text{(IV,19a)}$$

Wenn man Viscositätsmessungen auswerten möchte, die in einem anderen als den bei Gl. IV,17—19 angegebenen Lösungsmitteln oder bei einer anderen Temperatur durchgeführt wurden, so kann man die entsprechende Eichgleichung leicht mit Hilfe einiger Viscositätsmessungen aufstellen. Man mißt die Viscositätszahl einiger Polymerfraktionen im interessierenden Molekulargewichtsbereich zunächst unter den bei der originalen Eichgleichung angegebenen Bedingungen ($\rightarrow \bar{P}_\eta$) und dann in dem betreffenden Lösungsmittel bei der gewünschten Temperatur. Auf diese Weise erhielten wir z. B. für Polystyrol ($[\eta]$ bei 30° in Benzol gemessen):

$$[\eta] = 0{,}341\ \bar{P}_\eta^{0{,}72} \qquad \text{(IV,17a)}$$

für Polystyrol ($[\eta]$ bei 20° in Toluol gemessen):

$$[\eta] = 0{,}324\ \bar{P}_\eta^{0{,}72} \qquad \text{(IV,17b)}$$

und für Polymethylmethacrylat ($[\eta]$ bei 35° in Benzol gemessen):

$$[\eta] = 0{,}189\ \bar{P}_\eta^{0{,}76}\,. \qquad \text{(IV,18a)}$$

V. Die Kettenübertragung bei der Polymerisation verschiedener Monomerer

In diesem Abschnitt wird zunächst die Bestimmung von Übertragungskonstanten an einer Reihe von Beispielen beschrieben, wobei wir uns jedoch auf die Monomeren Styrol, Methylmethacrylat und Vinylacetat beschränken. Die hierbei verwendeten Meßdaten sind großenteils der Literatur entnommen. Die ausführliche Behandlung soll die Anwendung der in Abschnitt III abgeleiteten Gleichungen demonstrieren.

Anschließend geben wir eine tabellarische Zusammenstellung von bisher bekannten und nach den Gesichtspunkten der vorliegenden Arbeit als zuverlässig zu betrachtenden Übertragungskonstanten bei der Polymerisation der drei genannten Monomeren.

a) Beispiele von Kettenübertragung bei der Polymerisation von Styrol

Nach den bisherigen Kenntnissen erfolgt der Kettenabbruch bei Polystyrol ausschließlich durch Kombination (*19, 60*). Da ferner in allen hier verwendeten Arbeiten die Bestimmung der Polymerisationsgrade auf dem Weg über die Viscositätszahl erfolgte, wird zur Auswertung die durch Tab. 5, Zeile b, gegebene Gleichung

$$\frac{A_\eta}{\overline{P}_\eta} - \frac{k_{ab}\, v_{Br}}{k_w^2\,[M]^2} = \Sigma \frac{v_{ü}}{v_w} \tag{V,1}$$

verwendet.

A_η wird aus Abb. 2 (S. 515) entnommen, wobei der Abszissenwert $\overline{P}_\eta / \langle \overline{P}_{\eta,0} \rangle$ berechnet wird unter Verwendung des Ausdruckes für $\langle \overline{P}_{\eta,0} \rangle$ aus Tab. 5, Zeile b, und unter Berücksichtigung von $a = 0{,}72$ (siehe Gl. IV,17):

$$\frac{\overline{P}_\eta}{\overline{P}_{\eta,0}} = \frac{\overline{P}_\eta}{2{,}86} \cdot \frac{v_{Br} \cdot k_{ab}}{[M]^2 k_w^2} \tag{V,2}$$

Der Wert $\overline{P}_\eta / \langle \overline{P}_{\eta,0} \rangle$ ist direkt ein Maß für die Veränderung der Molekulargewichtsverteilung durch die Übertragung. $\overline{P}_\eta / \langle \overline{P}_{\eta,0} \rangle = 1$ bedeutet, daß alle Moleküle durch Kombinationsabbruch gebildet sind, d. h. aus zwei Polymerketten bestehen. $\overline{P}_\eta / \langle \overline{P}_{\eta,0} \rangle \to 0$ bedeutet, daß praktisch alle Moleküle durch Übertragung beendet sind und somit nur aus einer Polymerkette bestehen.

Das Viscositätsmittel des Polymerisationsgrades, $\overline{P}_\eta$, erhalten wir aus der experimentellen Viscositätszahl mit Hilfe von Gl. IV,17 oder einer danach aufgestellten Eichkurve bzw. mit einer der abgewandelten Gleichungen IV,17a oder IV,17b, je nachdem in welchem Lösungsmittel und bei welcher Temperatur die Viscositätszahl gemessen wurde.

Das Verhältnis der Geschwindigkeitskonstanten, k_w^2/k_{ab}, welches zur Auswertung von Gl. V,1 benötigt wird, berechnet man nach Gl. IV,12.

Die Auswertung von Gl. V,1 liefert in jedem Fall $\Sigma(v_{ü}/v_w)$, woraus dann, unter sinngemäßer Anwendung von Gl. III,64, die einzelnen Übertragungskonstanten bestimmt werden.

1. Kettenübertragung durch das Monomere. Bei der thermischen Polymerisation in Substanz ist die einzig mögliche Kettenübertragung jene durch das Monomere selbst, abgesehen von der Übertragung durch das Polymere, welche sich jedoch immer dadurch vermeiden läßt, daß man nur bis zu geringem Umsatz polymerisiert ($< 5\%$). Vom theoretischen Standpunkt aus ist die Übertragungskonstante des Monomeren daher bei dieser Art von Polymerisation am störungsfreisten zu erhalten. Praktisch ist jedoch zu berücksichtigen, daß die thermische Polymerisation außerordentlich empfindlich gegen Verunreinigungen bzw. Luftsauerstoff ist. Dadurch erklärt sich eine verhältnismäßig breite Streuung zwischen Messungen verschiedener Autoren.

Bei der thermischen Polymerisation in Substanz liefert die Auswertung nach Gl. V,1 direkt die Übertragungskonstante des Monomeren (vgl. III,64):

$$\Sigma \frac{v_{ü}}{v_w} = C_M \,. \tag{V,3}$$

Tabelle 7. *Übertragungskonstante des Monomeren bei der thermischen Substanzpolymerisation von Styrol*

T °C	$[M]$ Mol/l	$v_{Br} \cdot 10^6$ Mol/l sec	$\overline{P}_\eta$	$\frac{\overline{P}_\eta}{\langle \overline{P}_{\eta,0} \rangle}$ n. Gl. V, 2	$\frac{k_w^2}{k_{ab}} \cdot 10^4$ n. Gl. IV,12	$C_M \cdot 10^4$ n. Gl. V,1 und 3	Zitat
27	8,62	0,15	41300	0,29	1,0	0,31	*52*
50	8,47	0,82	23800	0,24	4,0	0,62*	*61*
60	8,37	2,00	18000	0,27	6,76	0,79*	*83*
70	8,28	5,08	10750	0,25	11,0	1,35	*52*
70	8,28	4,37	12500	0,25	11,0	1,16	*22*
90	8,12	22,6	7400	0,32	27,6	1,79	*26*
90	8,12	20,1	8800	0,34	27,6	1,47	*61*
100	8,03	49,0	6400	0,41	41,7	1,83	*126*
100	8,03	49,9	6580	0,42	41,7	1,72	*49*
110	7,25	106	4250	0,39	63,0	2,80	*26*
132	7,75	433	3300	0,59	140	2,45	*126*
132	7,75	472	2720	0,53	140	3,4	*49*
132	7,75	466	2160	0,41	140	5,33	*52*

* Mittelwerte aus einer Reihe von Messungen.

In Tab. 7 sind aus der Literatur entnommene Daten der thermischen Polymerisation von Styrol zusammengestellt, die auf diese Weise ausgewertet wurden.

In Tab. 8 sind noch einige weitere Werte für C_M aufgeführt, welche teils bei der thermischen Polymerisation in Lösung, teils bei der katalysierten Polymerisation gewonnen wurden. Der Hinweis in der letzten Spalte gibt an, in welchem der folgenden Abschnitte die Bestimmung besprochen wird.

Tabelle 8. *Übertragungskonstante des Monomeren, berechnet aus Daten der thermischen Lösungs- sowie der katalysierten Polymerisation (Styrol)*

Art der Polymerisation	T °C	$C_M \cdot 10^4$	Zitat
Thermisch, Lösung	60	0,85	V a 2
Thermisch, Lösung	60	0,8	V a 2
Thermisch, Lösung	100	1,8	V a 2
Thermisch, Lösung	100	1,7	V a 2
Thermisch, Lösung	132	3,0	V a 2
Katalysiert, Substanz	50	0,6	V a 3
Katalysiert, Substanz	60	0,6	V a 3
Katalysiert, Lösung	90	1,25	V a 3

In Abb. 5 sind alle Werte von C_M aus Tab. 7 und 8 als Funktion des Reziprokwertes der absoluten Temperatur dargestellt. In dem untersuchten Bereich läßt sich die Temperaturabhängigkeit der Übertragungskonstante des Monomeren am besten wiedergeben durch die Gleichung:

$$\log_{10} C_M = -0{,}87 - 1075/T\,. \qquad \text{(V,4)}$$

Dies entspricht der durchgezogenen Geraden. Hierbei wurden am stärksten die von MAYO u. Mitarb. (*49, 50*) aus der thermischen Lösungspolymerisation gewonnenen Punkte berücksichtigt, da diese für am zuverlässigsten gehalten werden (vgl. nächsten Abschnitt).

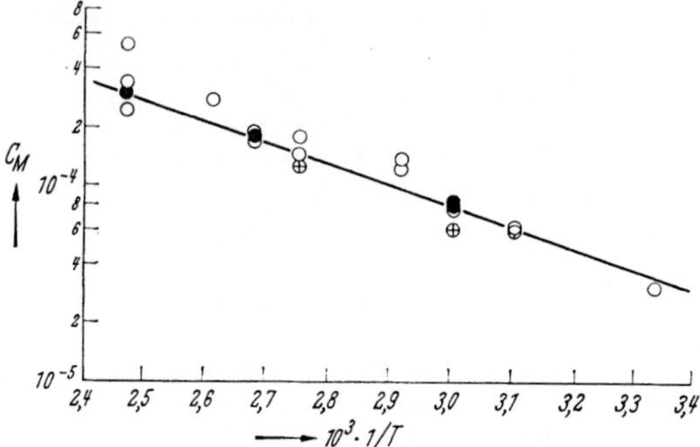

Abb. 5. Temperaturabhängigkeit der Übertragungskonstante des Monomeren bei der Polymerisation von Styrol. ○: aus thermischer Substanzpolymerisation; ●: aus thermischer Polymerisation in Lösung; ⊕: aus katalysierter Polymerisation

Aus Gl. V,4 läßt sich die Differenz der Aktivierungsenergien von Übertragung und Wachstum ausrechnen; sie beträgt $E_{ü} - E_w \cong 5$ kcal. Mit der Aktivierungsenergie des Kettenwachstums, welche man aus Gl. IV,12a erhalten kann ($E_w \cong 6$ kcal) ergibt sich:

$$E_{ü} \cong 11 \text{ kcal}\,.$$

Tabelle 9. *Messungen von* GREGG *und* MAYO *an Äthylbenzol,* ausgewertet nach Gl. V,1

$[M]$ Mol/l	$\frac{[LM]}{[M]}$	$v_{Br} \cdot 10^6$ Mol/l sec	$\bar{P}_\eta$	$\frac{\bar{P}_\eta}{\langle \bar{P}_{\eta,0} \rangle}$ n. Gl.V,2	$\Sigma \frac{v_{ü}}{v_w} \cdot 10^4$ n. Gl.V,1	$C_{LM} \cdot 10^4$ aus Abb. 6—8
$T = 60°$ C; $k_w^2/k_{ab} = 6{,}76 \cdot 10^{-4}$						
4,80	0,70	0,685	11260	0,17	1,4	0,83
2,79	1,88	0,244	7790	0,13	2,2	
2,68	1,99	0,250	6560	0,12	2,6	
2,31	2,47	0,154	5860	0,09	3,0	
1,85	3,32	0,134	4540	0,09	3,8	
0,64	11,4	0,023	1570	0,05	11,4	
$T = 100°$ C; $k_w^2/k_{ab} = 4{,}17 \cdot 10^{-3}$						
5,35	0,47	21,9	4800	0,31	2,8	2,2
3,76	1,06	10,5	3600	0,22	4,2	
2,63	1,92	5,66	2790	0,19	5,6	
1,52	4,01	1,96	1590	0,11	10,6	
0,75	9,10	0,42	790	0,05	22,5	
$T = 132°$ C; $k_w^2/k_{ab} = 1{,}4 \cdot 10^{-2}$						
4,31	0,75	118,7	2050	0,33	6,4	4,9
3,17	1,36	64,4	1580	0,25	9,2	
1,76	3,15	19,0	920	0,14	18,0	

2. Kettenübertragung durch das Lösungsmittel. Wenn in Lösung polymerisiert wird, kann außer dem Monomeren auch das Lösungsmittel ein Kettenüberträger sein. Bei der Auswertung nach Gl. V,1 erhält man dann (vgl. III,64):

$$\Sigma \frac{v_ü}{v_w} = C_M + C_{LM} \frac{[LM]}{[M]}. \qquad (V,5)$$

Wenn man also die aus Einzelversuchen berechneten Werte für $\Sigma(v_ü/v_w)$ gegen $[LM]/[M]$ graphisch aufträgt, erhält man als Ordinatenabschnitt die Übertragungskonstante des Monomeren, als Steigung jene des Lösungsmittels.

Über die übertragende Wirkung verschiedener Lösungsmittel bei der thermischen Polymerisation von Styrol liegen zahlreiche, sehr sorgfältige Messungen von GREGG und MAYO (*49*) vor, welche hier als Beispiel zur Auswertung nach V,1 und V,5 verwendet werden sollen.

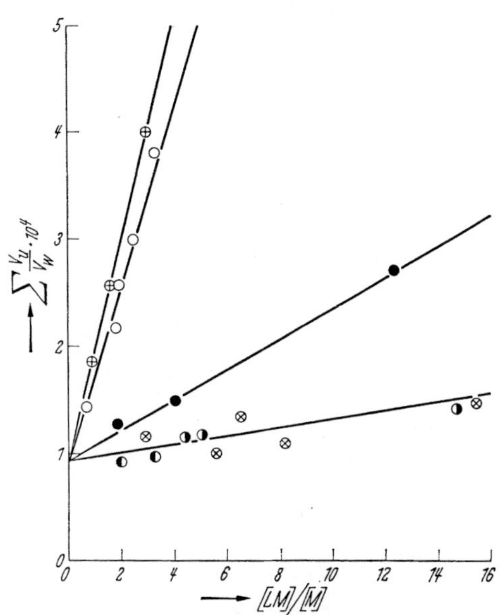

Abb. 6. Meßwerte von GREGG und MAYO (*49*) bei 60° C, dargestellt nach Gl. V,5. ⊗ Benzol; ◑ Cyclohexan; ◐ t-Butylbenzol; ● Toluol; ○ Äthylbenzol; ⊕ Isopropylbenzol

In Tab. 9 sind die Messungen von GREGG und MAYO an Äthylbenzol bei drei verschiedenen Temperaturen zusammengestellt und ausgewertet. Die Messungen an einer Reihe von weiteren Lösungsmitteln wurden in derselben Weise bearbeitet und sind in den Abb. 6—8 graphisch dargestellt.

Aus den Abbildungen kann die Übertragungskonstante des Monomeren als Ordinatenabschnitt direkt abgelesen werden. Da letzterer, für jede Temperatur, durch ein Büschel von geraden Linien festgelegt ist, erscheint diese Bestimmung der Übertragungskonstante des Monomeren als besonders zuverlässig (vgl. Abb. 5).

Die Übertragungskonstanten der Lösungsmittel, welche sich aus den Abbildungen als Steigung ergeben, sind in Tab. 10 zusammengestellt. In Klammern stehen jeweils die von MAYO nach Gl. I,2 errechneten Werte. Aus der Tabelle ist zu ersehen, daß die Abweichung (zumindest bei 60 und 100° C) nicht sehr groß, aber systematisch ist. Die Anwendbarkeit von Gl. I,2 ist, wie bereits früher (Abschnitt I, a) erwähnt, daran gebunden, daß die Bruttogeschwindigkeit der thermischen Polymerisation streng proportional dem Quadrat der Monomerkonzentration ist, und ferner, daß das Zahlenmittel des Polymerisationsgrades bekannt ist.

Die Temperaturabhängigkeit der Übertragungskonstanten ist in Abb. 9 graphisch dargestellt. In diese Darstellung sind Daten von G. V. SCHULZ u. Mitarb. (*126*) mit aufgenommen, die in derselben Weise ausgewertet wurden. Die Übereinstimmung zwischen den Werten der beiden verschiedenen Autoren ist verhältnismäßig gut.

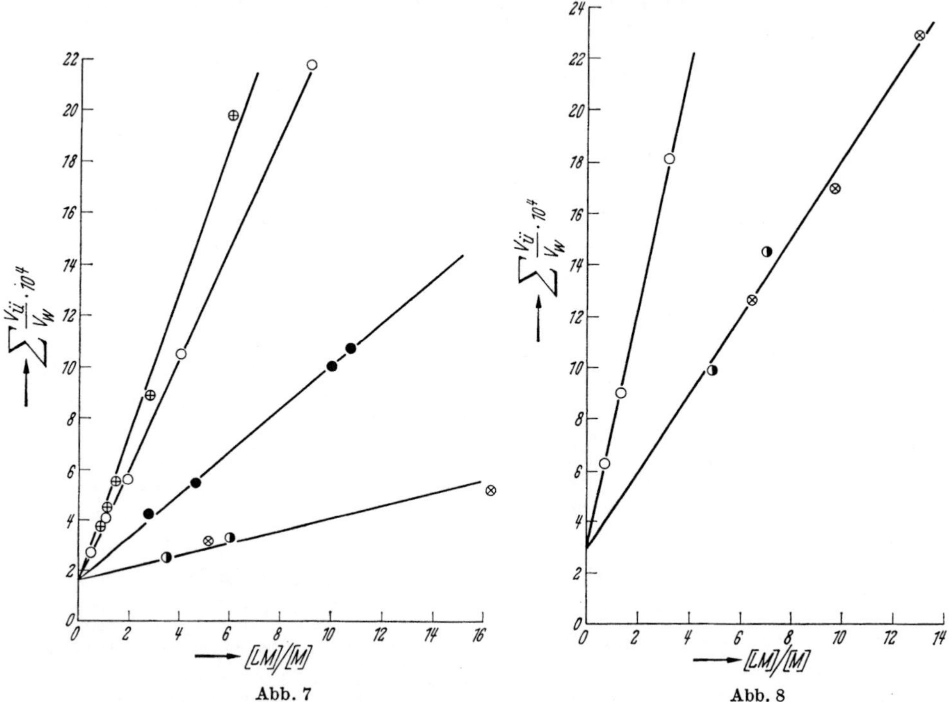

Abb. 7 Abb. 8

Abb. 7. Meßwerte von GREGG und MAYO bei 100° C, dargestellt nach Gl. V,5 (Zeichen wie in Abb. 6)
Abb. 8. Meßwerte von GREGG und MAYO bei 132° C, dargestellt nach Gl. V,5 (Zeichen wie in Abb. 6)

Tabelle 10.
Übertragungskonstanten von Lösungsmitteln bei der Polymerisation von Styrol
[berechnet aus Messungen von GREGG und MAYO (*49*)]

Lösungsmittel	$C_{LM} \cdot 10^5$			$E_ü - E_w$
	60 °C	100 °C	132 °C	
Benzol	0,4 (0,2)	2,3 (1,8)	15 (8,9)	13,4 (14,8)
Cyclohexan	0,4 (0,2)	2,3 (6,4)	15 (8,7)	13,4 (13,4)
t-Butylbenzol	0,4 (0,6)	—	—	—
Toluol	1,6 (1,2)	8 (6,4)	—	9,6 (10,1)
Äthylbenzol	8,3 (6,7)	22 (16,2)	49 (29)	6,6 (5,5)
Isopropylbenzol	10,4 (8,2)	29 (20)	—	6,1 (5,5)

Einige der von GREGG und MAYO in derselben Arbeit untersuchten Substanzen verursachen eine deutliche Beschleunigung der Polymerisation; z. B. Di- und Triphenylmethan, und, besonders ausgeprägt, Pentaphenyläthan. Die Autoren erklären dies durch einen zusätzlichen Kettenstart durch peroxydische Verunreinigungen oder durch eine leichte Dissoziation dieser Substanzen in Radikale. Aber auch dann liefert die Auswertung nach Gl. V,1 die richtigen Werte für die Übertragungskonstanten, da diese Gleichung von der Art des Kettenstartes völlig unabhängig ist. In Abb. 10 sind die Messungen an diesen drei Substanzen bei 60° graphisch dargestellt.

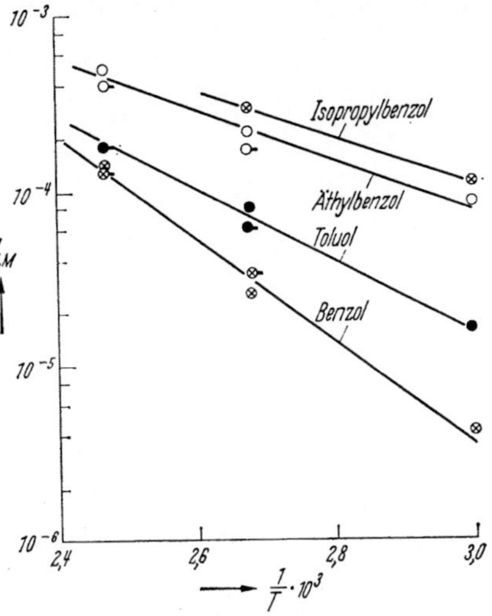

Abb. 9. Temperaturabhängigkeit der Übertragungskonstanten. ○: GREGG und MAYO (*49*); ○-: SCHULZ u. Mitarb. (*126*)

Man wird sich allerdings überlegen müssen, ob die auf diese Weise bestimmten Übertragungskonstanten in jedem Fall eindeutig sind, besonders dann, wenn es sich um aromatische Verbindungen handelt. Es ist seit langem bekannt (*151*), daß Radikale mit Aromaten auch unter Öffnung einer Doppelbindung reagieren können. Durch Verwendung von C^{14}-markiertem Benzol konnten STOCKMAYER u. Mitarb. (*141, 104*) zeigen, daß bei Polyvinylacetat mehrere Benzolmoleküle pro Polymerkette eingebaut werden. BURNETT u. Mitarb. (*3*) fanden ein ähnliches Resultat bei Polymethylmethacrylat. YOSHIDA (*153*) fand, bei Verwendung von

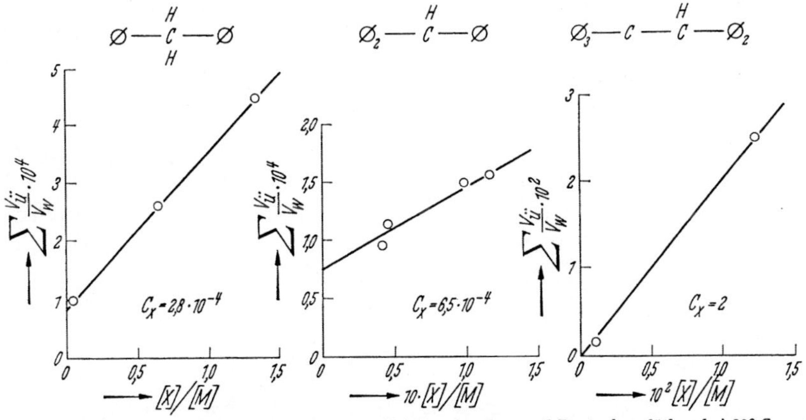

Abb. 10. Kettenübertragung von Diphenylmethan, Triphenylmethan und Pentaphenyläthan bei 60° C, nach Messungen von GREGG und MAYO (*49*), dargestellt nach Gl. V,5

deuteriertem Benzol als Lösungsmittel, zwar kein Deuterium im Polystyrol, aber CHITANI u. Mitarb. (*37*) berichten von ähnlichen Versuchen mit deuteriertem p-Diäthylbenzol, von welchem ebenfalls mehrere Moleküle pro Polymerkette eingebaut wurden. In aromatischen Lösungsmitteln ist also mit der Reaktion

$$P^* + \langle\rangle \rightarrow P-\langle\rangle^*$$

zu rechnen. Die dabei entstehenden Radikale können reaktionsträger als die wachsende Kette sein und liefern dann einen zusätzlichen Kettenabbruch. Die Übertragungskonstante könnte so ganz oder teilweise „vorgetäuscht" sein (vgl. Abschnitt III, e 1). Es wäre dann allerdings auch eine Verlangsamung der Geschwindigkeit zu erwarten. Über den thermischen Start der Polymerisation ist jedoch noch so wenig bekannt, daß es schwierig sein dürfte, zu definieren, wann die thermische Polymerisation „unverzögert" ist. [BREITENBACH u. Mitarb. (*23*) sowie PATAT und KIRCHNER (*71*, *102*) nehmen an, daß der thermische Start nach der 2. Ordnung bezüglich des Monomeren verläuft, wogegen MAYO (*80*, *82*) Start nach der 3. Ordnung vermutet.]

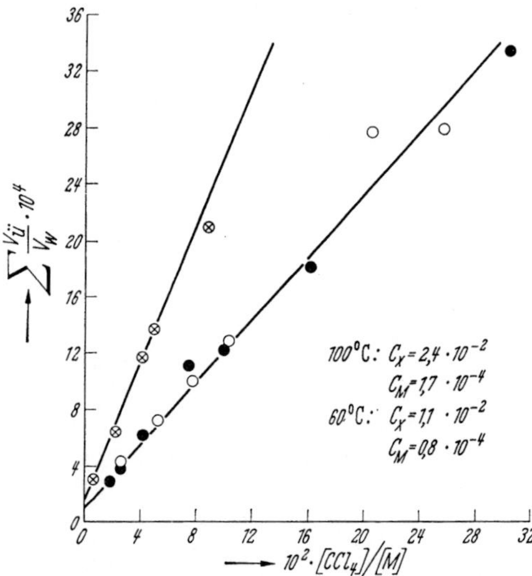

Abb. 11. Kettenübertragung durch Tetrachlorkohlenstoff [Meßwerte von GREGG und MAYO (*50*)]. Auftragung nach Gl. V,5. ⊗: thermisch 100° C; ●: thermisch 60° C; ○: katalysiert mit Bz_2O_2, 60° C ($[I]/[M] = 7 \cdot 10^{-5}$ Mol/l)

Bei Substanzen, die keine Doppelbindungen aufweisen, dürfte der Übertragungseffekt jedoch eindeutig sein. Abb. 11 zeigt Messungen von GREGG und MAYO (*50*) an Tetrachlorkohlenstoff. Diese Messungen sind besonders deswegen interessant, weil sie zeigen, daß man auch aus der katalysierten Polymerisation Übertragungskonstanten bestimmen kann, welche in guter Übereinstimmung mit denjenigen aus der thermischen Polymerisation sind. Bei Gegenwart eines Initiators muß berücksichtigt werden, daß auch dieser eine Kettenübertragung verursachen kann (siehe nächsten Abschnitt).

Es ist also:

$$\Sigma \frac{v_{\ddot{u}}}{v_w} = C_M + C_I \frac{[I]}{[M]} + C_{LM} \frac{[LM]}{[M]} . \qquad \text{(V,6)}$$

Man kann jedoch $[I]/[M]$ konstant halten und hat dann die ersten beiden Summanden rechts im Ordinatenabschnitt. Im vorliegenden

Fall ist $C_I[I]/[M]$ so niedrig, daß es gegenüber C_M nicht ins Gewicht fällt, daher liegen die Werte aus thermischer und katalysierter Polymerisation bei 60° auf derselben Geraden. Andernfalls wären parallele Geraden mit der Steigung C_{LM} zu erwarten (siehe z. B. Abb. 23).

In Abb. 12 sind eigene Messungen (*61*) zur Bestimmung der Übertragungskonstante von Dioxan dargestellt. Aus der Lage der Meßpunkte der katalysierten im Vergleich zu jenen der thermischen Polymerisation läßt sich abschätzen, daß $C_I[I]/[M]$ hier gegenüber C_M nicht ins Gewicht fällt.

Die Bestimmung von Übertragungskonstanten bei der katalysierten Polymerisation ist jedoch nicht generell möglich. In sehr vielen Fällen treten Nebenreaktionen auf, die die Auswertung unmöglich machen. In Abschnitt V, a 4 wird davon ausführlich die Rede sein.

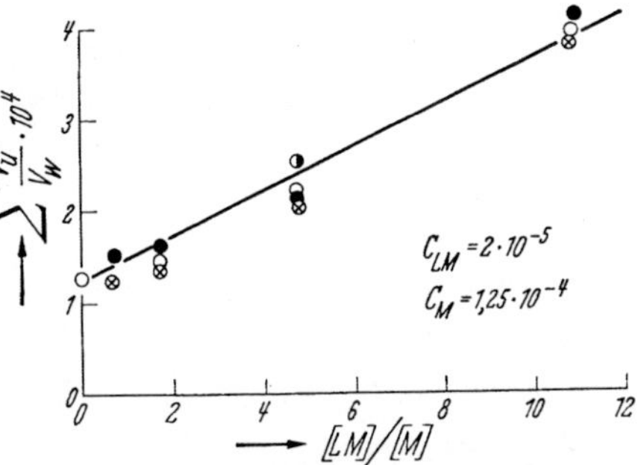

Abb. 12. Kettenübertragung von Dioxan gegenüber Polystyrol. $T = 90°$ C (*61*). ○: thermisch; ●: $[I]/[M] = 7{,}8 \cdot 10^{-4}$; ⊗: $[I]/[M] = 3{,}1 \cdot 10^{-3}$; ◐: $[I]/[M] = 1{,}9 \cdot 10^{-2}$ (I = t-Butylperoxyd)

3. Kettenübertragung durch Initiatoren. Übertragungskonstanten von Initiatoren werden am besten bei der Substanzpolymerisation bestimmt. Es ist dann:

$$\Sigma \frac{v_ü}{v_w} = C_M + C_I \frac{[I]}{[M]}. \quad \text{(V,7)}$$

Am ausführlichsten untersucht wurde der Initiator Benzoylperoxyd (Bz_2O_2). Messungen von Mayo u. Mitarb. (*83*) und von Baysal und Tobolsky (*12*) bei 60° C sowie eigene Messungen bei 50° C (*61*) wurden nach Gl. V,1 ausgewertet und in Abb. 13 gemäß Gl. V,7 graphisch dargestellt. Die außerordentlich hohe Übertragungskonstante ($C_I = 6\text{-}7 \cdot 10^{-2}$)

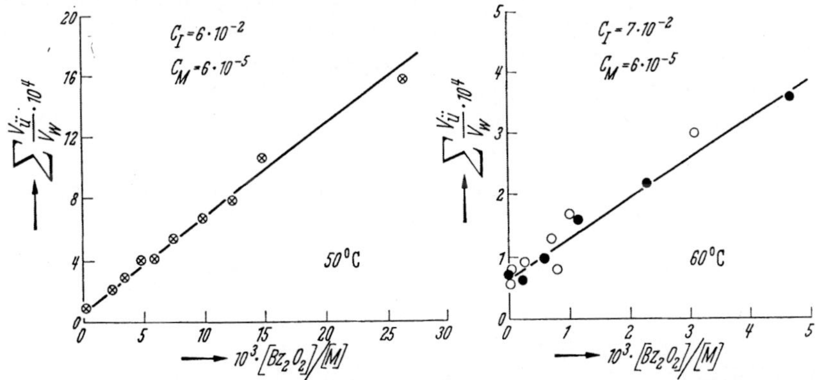

Abb. 13. Kettenübertragung durch Benzoylperoxyd, Auftragung nach Gl. V,7. ⊗: eigene Messungen bei 50° C; ●: Mayo u. Mitarb. bei 60° C (*83*); ○: Baysal und Tobolsky bei 60° C (*12*)

weist darauf hin, daß es sich hier nicht, wie in den meisten bisher besprochenen Fällen, um die Übertragung eines Wasserstoffatoms handeln kann. NOZAKI und BARTLETT (*96*) haben gezeigt, daß Benzoylperoxyd unter dem Einfluß von radikalliefernden Substanzen einen induzierten Zerfall erleidet. Die hohe Übertragungskonstante ist daher auf sekundären Zerfall des Bz_2O_2 unter der Einwirkung der Polymerradikale zurückzuführen, wobei letztere abgesättigt und neue Polymerketten gestartet werden.

Der zweite, sehr häufig verwendete Initiator ist Azobisisobutyronitril (AIBN). Bei dieser Substanz konnte kein sekundärer Zerfall festgestellt werden (*99, 151*). Es ist daher hier nur eine sehr geringe, den Methylgruppenwasserstoffen entsprechende Übertragungskonstante zu erwarten (vgl. Abschnitt V, d), welche jedoch wegen einer auftretenden Nebenreaktion nicht bestimmt werden kann. Hierüber wird im nächsten Abschnitt berichtet.

Prinzipiell ist die Bestimmung von C_I auch in Lösung möglich, wenn man $[LM]/[M]$ konstant hält und nach Gl. V,7 auswertet. Der Ordinatenabschnitt ist dann $C_M + C_{LM}[LM]/[M]$. Leider liegen bei Styrol keine geeigneten Messungen hierzu vor. In den meisten Fällen treten Nebenreaktionen auf, die diese Art der Bestimmung von C_I unmöglich machen.

4. Störungen durch Nebenreaktionen bei der katalysierten Polymerisation. Bei der Polymerisation unter Verwendung von radikalliefernden Initiatoren treten häufig Nebenreaktionen auf, die z. T. noch nicht ganz aufgeklärt sind und welche es in vielen Fällen nicht ratsam erscheinen lassen, diese Systeme zur Bestimmung von Übertragungskonstanten zu verwenden.

So liefert z. B. die formale Auswertung von Messungen der Substanzpolymerisation von Styrol mit AIBN nach Gl. V,1 eine „scheinbare Übertragungskonstante", welche fast ebenso groß ist wie diejenige von Bz_2O_2, nämlich $3 \cdot 10^{-2}$ (siehe Abb. 14). Wie bereits erwähnt, ist bei AIBN kein induzierter Zerfall bekannt, der eine so starke Übertragung erklären könnte. Wie kürzlich gezeigt wurde (*7a, 58*), ist dieser Effekt auf eine zusätzliche Abbruchsreaktion zurückzuführen, nämlich zwischen dem Initiator- und dem Polymerradikal.

Bei Berücksichtigung dieser Reaktion mit der Geschwindigkeit

$$v_{ab}''' = k_{ab}''' [R^*] [P^*] \qquad \text{(V,8)}$$

muß in Gl. V,1 ein zusätzliches Glied eingeführt werden:

$$\frac{A_\eta}{P_\eta} - B_\eta = \frac{k_{ab}''' [R^*]}{k_w [M]} + \Sigma \frac{v_{ü}}{v_w} ; \qquad \text{(V,9)}$$

wenn man den ersten Term rechts mit Gl. I,10:

$$v_{st} = k_{st} [R^*] [M] = f\, 2\, k_z [I]$$

erweitert, so erhält man anstelle von (V,1):

$$\frac{A_\eta}{\bar{P}_\eta} - B_\eta = C_M + \left(C_I + \frac{k'''_{ab} f 2 k_z}{k_{st} k_w [M]}\right) \frac{[I]}{[M]}. \qquad \text{(V,10)}$$

Da sich $f/[M]$ bei Substanzpolymerisation sehr wenig mit zunehmender Initiatorkonzentration ändert (*58*), ergibt sich die gerade Linie in Abb. 14.

Unter der Annahme, daß AIBN eine niedere, seiner Konstitution entsprechende Übertragungskonstante hat ($C_I < 10^{-4}$, vgl. z. B. Tab. 14), läßt sich aus der Steigung der Geraden in Abb. 14 das Verhältnis der Konstanten der beiden Reaktionen des Initiators abschätzen, nämlich:

$$\frac{k'''_{ab}}{k_{st}} \cong 1 \cdot 10^6 .$$

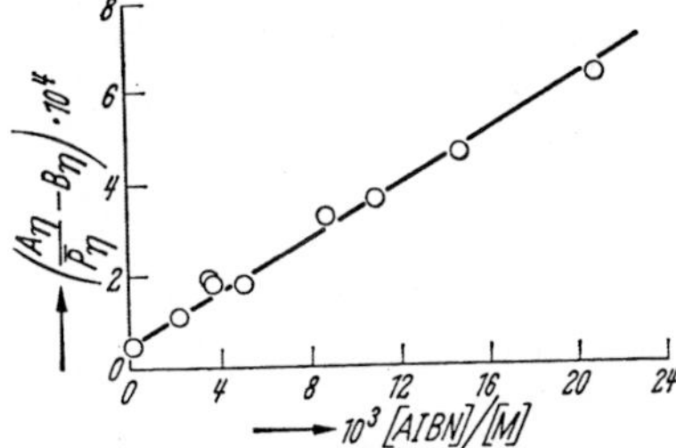

Abb. 14. Substanzpolymerisation von Styrol mit AIBN (*61*), dargestellt nach Gl. V,10. $T = 50°$ C

Diese Abbruchsreaktion tritt natürlich auch bei Lösungspolymerisation mit AIBN auf. Trotz Kenntnis von k'''_{ab}/k_{st} kann sie jedoch dort nicht immer quantitativ berücksichtigt werden, da das Verhältnis $f/[M]$ in Gl. V,10 von der Lösungsmittelkonzentration abhängt (*58*) und außerdem in verschiedenen Lösungsmitteln verschieden sein kann [vgl. z. B. die Beobachtungen von Burnett u. Mitarb. (*3, 28*) mit Brombenzol].

Inwieweit auch andere Initiatoren diesen Kettenabbruch liefern, ist bisher nicht untersucht worden.

Eine weitere, bei der Bestimmung von Übertragungskonstanten störende Nebenreaktion wurde zuerst bei der Polymerisation von Styrol mit AIBN in Benzol aufgefunden (*57, 58*) und inzwischen auch in anderen Lösungsmitteln mit aromatischem Charakter [z. B. Isopropylbenzol, Pyridin) beobachtet (*61*)]. Unter gleichen Bedingungen ist die Bruttogeschwindigkeit der Polymerisation in diesen Lösungsmitteln praktisch gleich wie in einem „inerten“ Lösungsmittel, z. B. Dioxan, jedoch werden die Polymerisationsgrade in den aromatischen Lösungsmitteln stets niederer gefunden, wobei der Unterschied mit zunehmender Lösungsmittelkonzentration zunimmt. Bei 20% Monomer ist z. B. der Polymerisationsgrad in Benzol 40% niederer als in Dioxan. Die Übertragungskonstante des Benzols müßte etwa 2 Größenordnungen höher sein, als bei der thermischen Polymerisation gefunden wurde, um diese Verminderung der Polymerisationsgrade zu erklären.

Der Mechanismus dieser Nebenreaktion konnte bisher nicht aufgeklärt werden. Möglicherweise handelt es sich um die Bildung eines Komplexes des Initiators mit dem aromatischen Lösungsmittel, welcher eine starke Überträgerwirkung hat, oder aber um Copolymerisation, wie sie bereits bei der thermischen Polymerisation als möglich erwähnt wurde (siehe Abschnitt V, a 2).

Wenn man Messungen der katalysierten Polymerisation von Styrol in einem aromatischen Lösungsmittel ohne Berücksichtigung dieser Nebenreaktion formal nach Gl. V,1 auswertet und nach Gl. V,6 graphisch darstellt, so erhält man Kurvensysteme von der Art der Abb. **15**. Es ist offensichtlich, daß man einen von $[I]/[M]$ abhängigen Wert für C_{LM} und einen von $[LM]/[M]$ abhängigen Wert für C_I erhält.

Es soll daher an dieser Stelle ausdrücklich davor gewarnt werden, Übertragungskonstanten bei der katalysierten Polymerisation zu be-

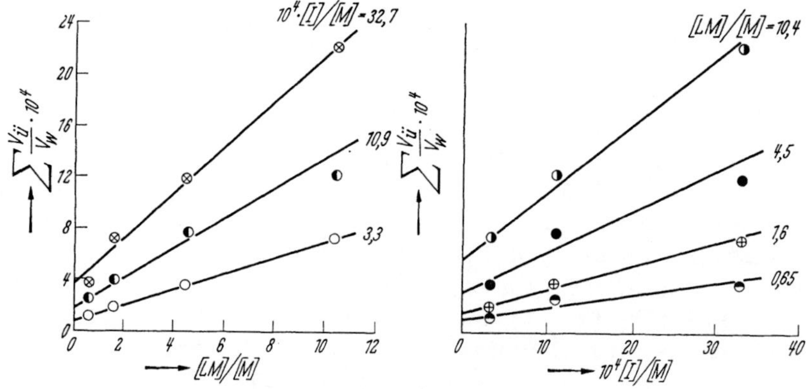

Abb. 15.
Polymerisation von Styrol mit AIBN in benzolischer Lösung (*61*). $T = 50°$ C. Darstellung nach Gl. V,6

stimmen, ohne den Nachweis erbracht zu haben, daß derartige störende Nebenreaktionen nicht auftreten.

Auf die Abwesenheit derartiger Nebenreaktionen kann geschlossen werden, wenn sich das Kurvensystem in Abb. **15** zu einer Kurve vereinigt (siehe z. B. Abb. **11** und **12**), oder aber, wenn sich parallele Geraden ergeben (siehe Abb. **23**).

5. Übertragung durch „Regler“. Als „Regler“ werden Substanzen mit besonders hohen Übertragungskonstanten bezeichnet, welche technisch zur Regulierung des Polymerisationsgrades von Interesse sind. Besonders Mercaptane haben sich hierfür als geeignet erwiesen. Ihre Übertragungskonstanten liegen in der Größenordnung 10^{-1}—10^{1}.

Bei derartig starken Effekten ist der Verbrauch an Überträgersubstanz (X) sehr hoch, und selbst bei niederen Umsätzen kann nicht mehr mit konstanter Konzentration $[X]$ gerechnet werden.

Es empfiehlt sich dann eine andere Methode, welche zuerst von W. V. SMITH (*137*) vorgeschlagen wurde.

Das Verhältnis der Geschwindigkeiten von Übertragung und Wachstum ist (vgl. I,11 und 12):

$$\frac{v_{\ddot{u},x}}{v_w} = -\frac{d[X]}{dt} \Big/ -\frac{d[M]}{dt} = C_x \frac{[X]}{[M]}.$$

Daraus folgt:

$$\frac{d \ln [X]}{d \ln [M]} = C_x. \qquad \text{(V,11)}$$

Die Übertragungskonstante kann also dadurch bestimmt werden, daß man den Verbrauch an Überträger und an Monomerem während der Polymerisation verfolgt. Im Falle der Mercaptane kann die jeweils noch in der Lösung befindliche Konzentration an Überträger z. B. durch amperometrische Titration (mit Silbernitrat) bestimmt werden. Eine andere sehr empfindliche Methode ist die Verwendung radioaktiv markierter Überträgersubstanzen [z. B. S^{35} bei Mercaptanen (*150*)]. Der Verbrauch des Monomeren wird am besten durch Ausfällen des Polymeren bestimmt. Auch die Veränderung des Brechungsindex kann als Maß herangezogen werden.

In Abb. 16 wird diese Art der Bestimmung von Übertragungskonstanten am Beispiel des Dodecylmercaptans, welches von MAYO u. Mitarb. (*48*) untersucht wurde, gezeigt. Die Übertragungskonstante ergibt sich aus der Neigung der Geraden zu 18,5.

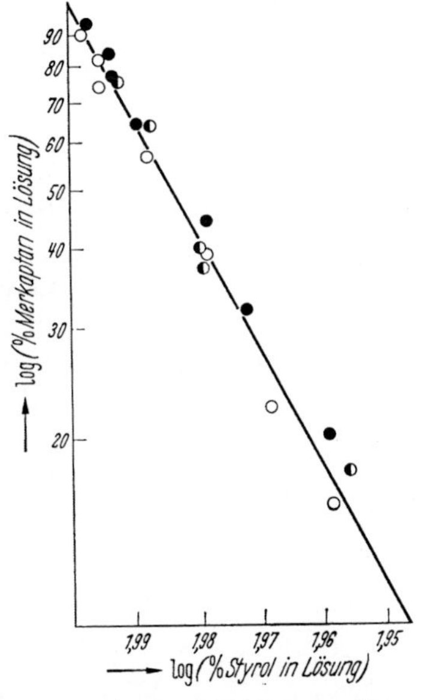

Abb. 16. Zur Bestimmung der Übertragungskonstante von Dodecylmercaptan bei Styrol, nach MAYO u. Mitarb. (*48*). $T = 60°$ C. (Mit freundlicher Genehmigung der Am. Chem. Soc.) ●: 0,311% Mercaptan; ◐: 1,99% Mercaptan; ○: 0,446% Mercaptan in benzolischer Lösung

6. Übertragung durch das Polymere. Wenn Kettenübertragung am Polymeren auftritt, so entstehen verzweigte[1] Polymere (*44*). Setzt man einem polymerisierenden System jedoch fremdes Polymeres zu, an welchem Kettenübertragung stattfindet, so erhält man Block- oder Graft-Copolymere, je nachdem, ob die Übertragung am Ende oder in der Mitte des zugesetzten Polymeren geschieht (*136*).

Den ersten Versuch, Geschwindigkeitskonstanten derartiger Übertragungen am Polymeren zu bestimmen, haben BEVINGTON, GUZMÁN und MELVILLE (*17, 18*) gemacht. Bei der Polymerisation von C^{14}-markiertem Styrol setzen sie inaktives Polystyrol zu, dessen Polymerisationsgrad

[1] Wenn die Endgruppe des Polymeren der Kettenüberträger ist, erfolgt keine Verzweigung, sondern nur eine Molekülverlängerung.

sehr viel größer ist als derjenige des neuen Polymerisates. Das gesamte Polymere wird ausgefällt, fraktioniert, und die höchsten Fraktionen werden auf radioaktive Seitenzweige untersucht. Aus der Zahl der Seitenzweige wird die Konstante berechnet. Da hierbei einige Annahmen gemacht werden (z. B. über die Länge der Seitenzweige und die Verteilung der Aktivität in den Fraktionen), sind die so erhaltenen Konstanten mit einer gewissen Unsicherheit behaftet.

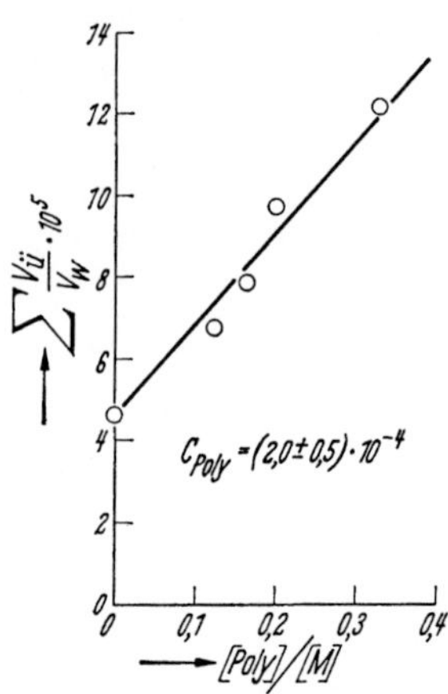

Abb. 17. Kettenübertragung durch das Polymere bei der Polymerisation von Styrol; $T = 50°$ C. Eigene Messungen (*63*), dargestellt nach Gl. V,12

Die Übertragungskonstante des Polymeren kann jedoch ebenfalls mit den in der vorliegenden Arbeit abgeleiteten Gleichungen bestimmt werden. Wenn man sehr niedermolekulares Polymeres (α-Polymeres) einem Ansatz zufügt, welcher zu einem hohen Polymerisationsgrad führt, so setzt das α-Polymere — analog niedermolekularen Überträgersubstanzen — den Polymerisationsgrad des entstehenden Polymeren (β-Polymeren) herab (*129, 62, 63*).

Morton und Piizma (*95*) haben versucht, das unverbrauchte α-Polymere vom neu entstandenen β-Polymeren quantitativ abzutrennen durch Fraktionieren, um so den Polymerisationsgrad des letzteren bestimmen zu können. Eigene Versuche mit diesem Verfahren haben uns auf Grund unbefriedigender Ergebnisse veranlaßt, dieses aufzugeben und die Viscositätszahl des β-Polymeren aus der Viscosität der gesamten Reaktionslösung auszurechnen (*129*).

Bei Gegenwart des α-Polymeren ist:

$$\Sigma \frac{v_{ü}}{v_w} = \left(\Sigma \frac{v_{ü}}{v_w}\right)_0 + C_{Poly} \frac{[Poly]}{[M]}. \qquad \text{(V,12)}$$

Den ersten Summanden rechts erhält man aus der Polymerisation ohne Zusatz des α-Polymeren unter sonst gleichen Bedingungen.

In Abb. 17 ist $\Sigma(v_{ü}/v_w)$, berechnet nach Gl. V,1, graphisch dargestellt nach Gl. V,12, für den Fall der Übertragung des Polystyrols. [Eigene Messungen (*63*).] Der Polymerisationsgrad des untersuchten α-Polystyrols ist 170, wohingegen die β-Polymerisation ohne Zusatz des α-Polymeren ein Polystyrol vom Polymerisationsgrad 7600 liefert.

Nach demselben Verfahren können natürlich auch die Übertragungskonstanten von fremden Polymeren (Pfropfkonstanten) bestimmt werden (*129, 63*).

7. Übertragung bei Verzögerung. Wenn gleichzeitig mit der Kettenübertragung eine Verzögerung der Bruttogeschwindigkeit auftritt, muß anstelle von Gl. V,1 die folgende Gleichung herangezogen werden, welche sich aus Tab. 5, Zeile e, entnehmen läßt:

$$\frac{A_\eta}{P_\eta} - \frac{k_{ab}\, v_{Br}}{k_w^2\,[M]^2}\left(\frac{v_{Br,0}}{v_{Br}}\right)^2 = \Sigma\,\frac{v_{ü}}{v_w}\,. \qquad \text{(V,13)}$$

A_η wird auch hier wieder mit Hilfe von V,2 und Abb. 2 bestimmt.

Eine Reihe von aromatischen Nitroverbindungen verzögern die Polymerisation von Styrol und setzen gleichzeitig den Polymerisationsgrad

Tabelle 11. *Kettenübertragung von* ($Bu_3P + Bu_2PH$) *gegenüber Polystyrol,* berechnet aus Messungen von E. PERRY (*106*). $T = 100°$ C; thermische Polymerisation

$\frac{[X]}{[M]}\cdot 10^2$	$v_{Br}\cdot 10^5$ Mol/l sec	$\frac{v_{Br}}{[M]^2}\cdot 10^7$	$\overline{P}_\eta$	$B_\eta\cdot 10^4$ n. Gl. V,10	$\frac{A_\eta}{\overline{P}_\eta}\cdot 10^4$	$\Sigma\frac{v_{ü}}{v_w}\cdot 10^4$
—	5,48	8,5	7200	2,04	3,4	1,4
0,3	4,35	6,8	1105	2,54	17,5	15,0
1,4	3,96	6,5	280	2,69	67,2	64,5

herab (*117, 20*). Da es sich hierbei jedoch wahrscheinlich nicht um Übertragungsreaktionen, sondern um den Angriff des Polymerradikals auf eine Doppelbindung handelt (*5, 147*), sollen diese Reaktionen hier nicht behandelt werden.

Einen typischen Fall von Verzögerung der Bruttogeschwindigkeit im Zusammenhang mit Kettenübertragung stellen die substituierten Phosphine dar, die von J. PELLON (*105*) und von E. PERRY (*106*) untersucht wurden.

In Tab. 11 sind Messungen von E. PERRY ausgewertet, welche an einer Mischung von Dibutylphosphin und Tributylphosphin, etwa im Verhältnis 1:4, durchgeführt wurden. Aus der dritten Spalte der Tabelle ist zu ersehen, daß eine Verzögerung der Geschwindigkeit von 20—25% auftritt. Ein Vergleich der 5. und 6. Spalte zeigt, daß $A_\eta/\overline{P}_\eta$ so groß ist gegenüber B_η, daß die Korrektur für die Verzögerung sich nur unwesentlich auswirkt.

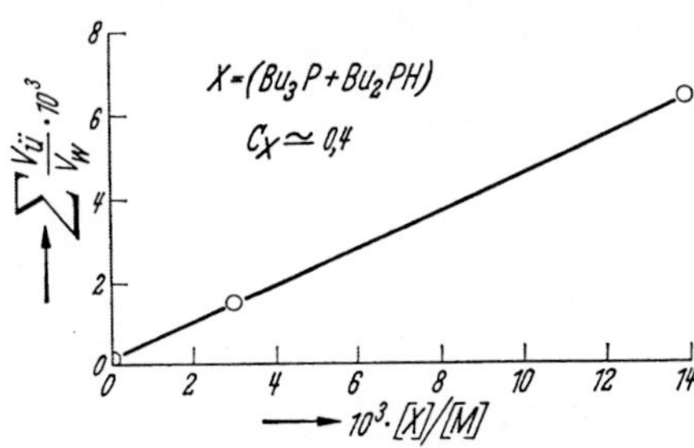

Abb. 18. Kettenübertragung von ($Bu_3P + Bu_2PH$) gegenüber Polystyrol. $T = 100°$ C. Messungen von E. PERRY (*106*), ausgewertet nach (V,13)

Dies dürfte allgemein der Fall sein bei Polymerradikalen, die eine relativ hohe Resonanzstabilisierung haben, wie z. B. Polystyrol. Eine Verzögerung tritt ja nur dann auf, wenn das bei der Übertragung entstehende Radikal X^* stabiler ist als die wachsende Polymerkette. Starke Resonanzstabilisierung des Radikals X^* bedingt aber eine relativ schwache Bindung XH, d. h. starken Übertragungseffekt. Wichtiger ist die Berücksichtigung der Verzögerung wahrscheinlich bei weniger resonanzstabilisierten Polymerradikalen, wie z. B. denjenigen des Polyvinylacetats. Leider liegen hierüber keine geeigneten Messungen vor.

Aus Abb. 18 ist zu entnehmen, daß die untersuchte 1:4-Mischung eine Übertragungskonstante von $\cong 0{,}4$ hat. Da die Konstante des Bu_3P 2 Größenordnungen niederer ist (siehe Tab. 17), ist jene des Bu_2PH als $\cong 2$ anzugeben.

b) Beispiele von Kettenübertragung bei der Polymerisation von Methylmethacrylat

Bei diesem Monomeren kommen die beiden Abbruchsreaktionen, Kombination und Disproportionierung, nebeneinander vor (siehe Abschnitt IV,a). Wegen der verschiedenen Aktivierungsenergien der beiden Reaktionen verschiebt sich das Verhältnis bei höheren Temperaturen zugunsten der Disproportionierung.

Da auch bei den hier in Betracht gezogenen Arbeiten die Molekulargewichtsbestimmung auf dem Weg über die Viscositätszahl erfolgte, wird zur Auswertung die durch Tab. 5, Zeile c, gegebene Gleichung herangezogen:

$$\frac{A_\eta}{\overline{P}_\eta} - \frac{k_{ab}\, v_{Br}}{k_w^2\, [M]^2} = \Sigma\, \frac{v_{ü}}{v_w}\,. \tag{V,14}$$

A_η wird auch hier aus Abb. 2 (S. 515) entnommen, wobei der Abszissenwert $\overline{P}_\eta / \langle \overline{P}_{\eta,0} \rangle$ berechnet wird unter Verwendung des Ausdruckes für $\langle \overline{P}_{\eta,0} \rangle$ aus Tab. 5, Zeile c, und unter Berücksichtigung von $a = 0{,}76$ (siehe Gl. IV,18):

$$\frac{\overline{P}_\eta}{\langle \overline{P}_{\eta,0} \rangle} = \frac{\overline{P}_\eta}{2{,}88} \cdot \frac{k_{ab}\, v_{Br}}{k_w^2\, [M]^2} \cdot \frac{1}{\delta + 1}\,. \tag{V,15}$$

$\overline{P}_\eta$ wird nach (IV,18) bzw. (IV,18a) aus der Viscositätszahl ermittelt, k_w^2/k_{ab} nach (IV,13). Das Verhältnis der beiden Abbruchskonstanten, δ, erhält man nach Gl. IV,7.

Tabelle 12. *Übertragungskonstante des Monomeren bei der thermischen Polymerisation von Methylmethacrylat*

T °C	$[M]$ Mol/l	$v_{Br} \cdot 10^6$ Mol/l sec	$\overline{P}_\eta \cdot 10^{-3}$	$\frac{k_w^2}{k_{ab}} \cdot 10^2$ n. Gl. IV, 13	δ n. Gl. IV, 7	$\frac{\overline{P}_\eta}{\langle \overline{P}_{\eta,0} \rangle}$ n. Gl. V, 15	$C_M \cdot 10^5$ n. Gl. V, 14
50	9,08	0,42	120	0,7	0,63	0,02	1,5
	9,08	0,39	127	0,7	0,63	0,02	1,4
60	8,96	1,15	98	1,03	0,79	0,03	1,8
	8,96	0,84	107	1,03	0,79	0,02	1,7
70	8,84	0,98	80	1,48	0,94	0,01	2,3
	8,84	1,35	81	1,48	0,94	0,02	2,2
80	8,74	3,08	72	2,15	1,15	0,02	2,4
	8,74	2,29	72	2,15	1,15	0,02	2,5
100	8,52	5,5	45	4,0	1,62	0,01	4,0
	8,52	5,1	50	4,0	1,62	0,01	3,6
120	8,30	14,0	33	7,1	2,19	0,01	5,4
	8,30	15,4	29	7,1	2,19	0,01	6,2

1. Kettenübertragung durch das Monomere. Wie bereits bei Styrol erwähnt, ist die thermische Polymerisation am besten zur Bestimmung der Übertragungskonstante des Monomeren geeignet. Im Gegensatz zum Styrol liegen jedoch bei Methylmethacrylat keine übereinstimmenden Messungen der thermischen Polymerisation in der Literatur vor (*6*).

Eigene Messungen (*61*) der thermischen Polymerisation in Substanz, mit speziell gereinigtem Methylmethacrylat, im Temperaturbereich von 50—120° C sind in Tab. 12 zusammengestellt und gemäß Gl. V,14 ausgewertet.

In Abb. 19 ist die Temperaturabhängigkeit der Übertragungskonstanten des Monomeren dargestellt. Ein Wert, der aus Messungen von GORNICK und O'BRIEN (*97*) bei der katalysierten Polymerisation berechnet wurde, paßt gut auf die Gerade (siehe nächsten Abschnitt). In die Abbildung ist auch die Temperaturabhängigkeit der Bruttogeschwindigkeit der thermischen Polymerisation mit aufgenommen. In dem untersuchten Bereich gilt:

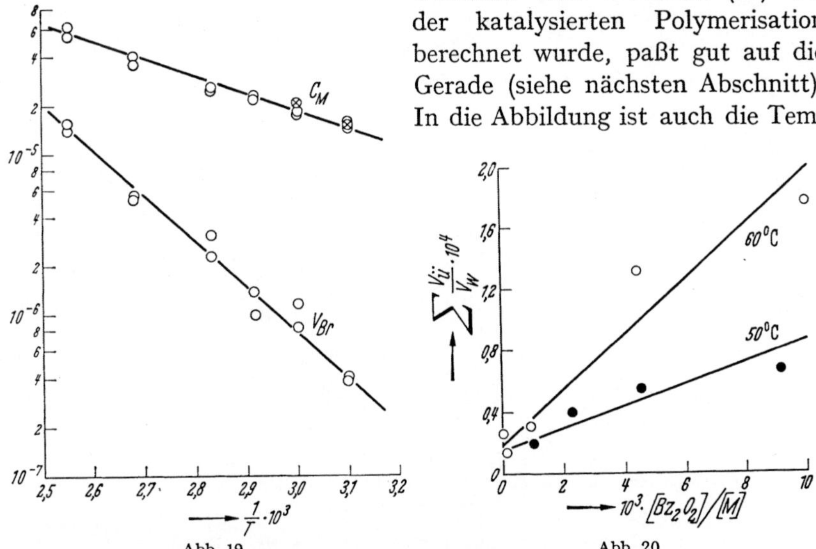

Abb. 19. Temperaturabhängigkeit der Übertragungskonstante des Monomeren sowie der thermischen Bruttogeschwindigkeit von Methylmethacrylat. ○: thermische Polymerisation in Substanz (*61*); ⊗: katalysierte Polymerisation in Substanz (s. Abb. 20 und 21)

Abb. 20. Kettenübertragung durch Benzoylperoxyd bei der Substanz-Polymerisation von Methylmethacrylat. Auftragung nach Gl. V,7. ●: SCHULZ und HARBORTH (*127*), 50° C; ○: BAYSAL und TOBOLSKY (*12*), 60° C

$$\log_{10} v_{Br,th} = 2{,}38 - 2840/T \qquad \text{(V,16)}$$

$$\log_{10} C_M = -1{,}55 - 1060/T \qquad \text{(V,17)}$$

2. Kettenübertragung durch Initiatoren. Der Initiator Benzoylperoxyd verhält sich bei der Polymerisation von Methylmethacrylat sehr ähnlich wie bei derjenigen von Styrol. Dies ist aus Abb. 20 zu ersehen, wo Messungen von SCHULZ und HARBORTH (*127*) bei 50° C und solche von BAYSAL und TOBOLSKY (*12*) bei 60° C gemäß Gl. V,7 dargestellt sind. Die Berechnung der $\Sigma(v_{ü}/v_w)$ erfolgte nach Gl. V,14. Aus der Steigung der Geraden ergibt sich $C_I \cong 0{,}01$ (50°C) bzw. $C_I \cong 0{,}02$ (60° C).

Messungen mit dem Initiator AIBN von O'BRIEN und GORNICK (*97*), ausgewertet nach Gl. V,14, sind in Abb. 21 wiedergegeben. Bei niederen Initiatorkonzentrationen liefern diese Messungen die Übertragungskonstante des Monomeren (praktisch keine Steigung; vgl. Gl. V,7). Bei höheren Initiatorkonzentrationen tritt ein plötzlicher Anstieg ein. Ob sich hier ebenfalls, wie bei Styrol, ein Abbruch durch Primärradikale andeutet, kann nicht entschieden werden, da nicht genügend Messungen vorliegen.

3. Kettenübertragung durch Lösungsmittel. Die thermische Polymerisation von Methylmethacrylat in verschiedenen Lösungsmitteln wurde von PALIT u. Mitarb. (*10*)

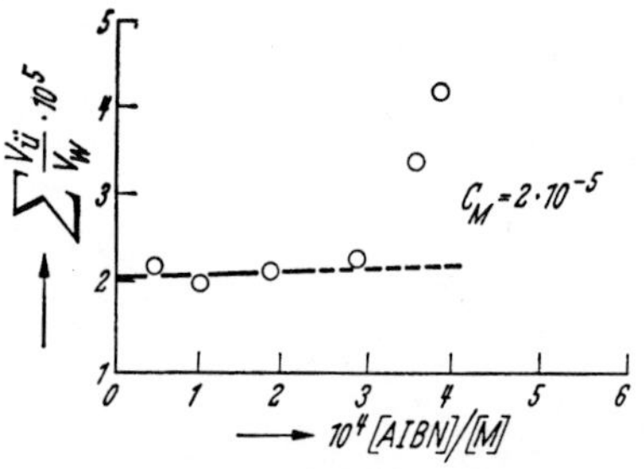

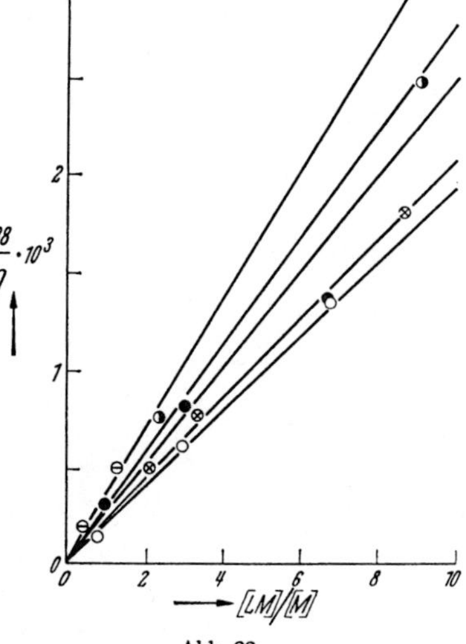

Abb. 21 Abb. 22

Abb. 21. Substanzpolymerisation von Methylmethacrylat mit AIBN. Meßwerte von O'BRIEN u. GORNICK (*97*), dargestellt nach Gl. V,7. $T = 60°$ C

Abb. 22. Kettenübertragung von Lösungsmitteln bei der thermischen Polymerisation von Methylmethacrylat. Messungen von PALIT u. Mitarb. (*10*). Aufgetragen nach (V,18). $T = 80°$ C. ⊖: CCl_4; ◑: Diäthylketon; ●: Isopropylbenzol; ⊗: Äthylbenzol; ○: Chloroform

untersucht. Da in der zitierten Arbeit die Bruttogeschwindigkeiten nicht angegeben sind, können nur solche Versuche ausgewertet werden, bei denen die Übertragung so stark ist, daß der Kombinationsabbruch daneben vernachlässigt werden kann. Man kann dann mit der Gleichung für reinen Disproportionierungsabbruch auswerten (Tab. 5, Zeile a). Da bei starker Übertragungswirkung ferner $A_\eta/\overline{P}_\eta \gg B$, ergibt sich für solche Fälle als gute Näherung:

$$\frac{1{,}88}{\overline{P}_\eta} \cong C_M + C_{LM} \frac{[LM]}{[M]} . \qquad \text{(V,18)}$$

Meßergebnisse der Polymerisation in einigen Lösungsmitteln mit relativ hohen Übertragungskonstanten sind in Abb. 22 gemäß Gl. V,18 dargestellt. Der Ordinatenabschnitt liefert $C_M = 3 - 4{,}10^{-5}$, in guter Übereinstimmung mit Abb. 19. Die C_{LM} sind in Tab. 12a zusammengestellt. Mit aufgenommen sind die Übertragungskonstanten von Benzol und Toluol, welche aus zwei Einzelversuchen berechnet wurden, bei denen die Geschwindigkeit[1] angegeben ist. Die Auswertung erfolgte hier nach (V,14).

Bei der *katalysierten* Polymerisation bestimmten CHADHA, SHUKLA und MISRA (*36*) die Übertragungskonstanten einer Reihe von Lösungsmitteln. Die Auswertung erfolgte jedoch nach Gl. I,2 (mit $[I]/[M] = \text{const}$); die berechneten Konstanten bedürfen daher, wie in Abschnitt I a erläutert, mehrerer Korrekturen. Ob letztere einen wesentlichen Einfluß auf die Ergebnisse haben würden, kann leider nicht überprüft werden, da die Bruttogeschwindigkeiten nicht angegeben sind und somit eine Neuberechnung nach den in der vorliegenden Arbeit vorgeschlagenen Gleichungen nicht erfolgen kann. Vor allem müßte überprüft werden, ob die bei Styrol beobachteten Nebenreaktionen zwischen dem Initiator und dem Lösungsmittel nicht auch hier die Ergebnisse beeinflussen (vgl. Abschnitt V, a 4, besonders Abb. 15).

Tabelle 12a. *Übertragungskonstanten einiger Lösungsmittel bei der thermischen Polymerisation von Methylmethacrylat.* $T = 80°$ C. [Berechnet aus Messungen von PALIT u. Mitarb. (*10*)]

Lösungsmittel	$C_{LM} \cdot 10^5$
Benzol	2,4
Toluol	5,2
Chloroform	19
Äthylbenzol	21
Isopropylbenzol	24
Diäthylketon	27
Tetrachlorkohlenstoff	33

Einen Fall von reiner Kettenübertragung ohne störende Nebenreaktionen scheint das System Methylmethacrylat, Äthylacetat, O-Cl-Benzoylperoxyd darzustellen, das von PALIT u. Mitarb. (*101*) untersucht wurde. Da sowohl Monomeres als auch Lösungsmittel und Initiator übertragen, ist

$$\Sigma \frac{v_{ü}}{v_w} = C_M + C_{LM} \frac{[LM]}{[M]} + C_I \frac{[I]}{[M]}. \qquad \text{(V,19)}$$

Aus den insgesamt 8 Meßpunkten der zitierten Arbeit lassen sich 3 Serien mit konstantem $[LM]/[M]$ und 2 Serien mit konstantem $[I]/[M]$ zusammenstellen. Die nach Gl. V,14 berechneten $\Sigma(v_ü/v_w)$ sind in Abb. 23a gegen $[I]/[M]$, in Abb. 23b gegen $[LM]/[M]$ aufgetragen. Durch die

[1] Die von PALIT u. Mitarb. (*10*) bei $[LM]/[M] = 1$ beobachtete thermische Bruttogeschwindigkeit ist etwa 4 mal größer als die von uns bei derselben Temperatur in Substanz gemessene. Dies ist möglicherweise auf peroxydische Verunreinigungen im Methacrylat zurückzuführen. Da die Bestimmung von Übertragungskonstanten nach Gl. V,14 jedoch unabhängig von der jeweiligen Startgeschwindigkeit ist, kann C_{LM} trotzdem auch in diesen Fällen bestimmt werden, wobei allerdings angenommen werden muß, daß die peroxydische Verunreinigung selbst kein starker Überträger ist.

Meßpunkte lassen sich in beiden Fällen parallele Geraden ziehen, deren Steigung die Übertragungskonstanten des Initiators (a) und des Lösungsmittels (b) gibt. Die parallelen Geraden in der Darstellung a müssen so gelegt werden, daß ihre Ordinatenabschnitte sich als unterste Gerade ($[I]/[M] = 0$) in Darstellung b einpassen und umgekehrt. Dadurch ist das ganze Kurvensystem sehr festgelegt. Als Ordinatenabschnitt der beiden untersten Kurven erhält man C_M.

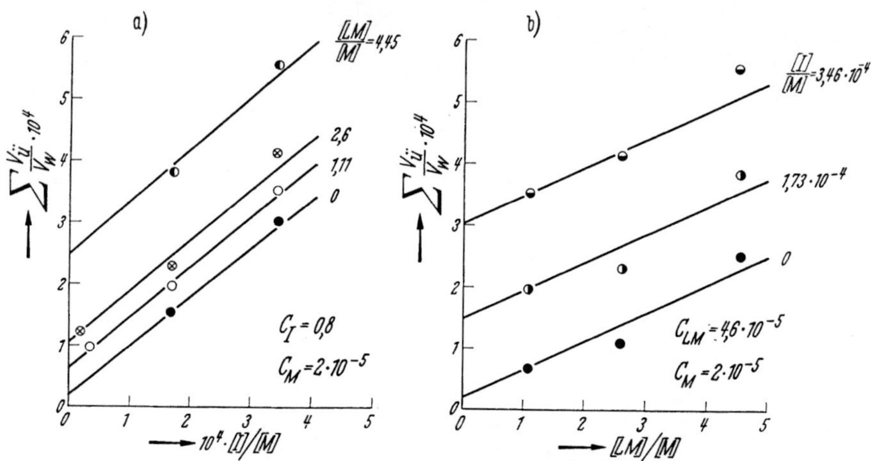

Abb. 23. Kettenübertragung des Lösungsmittels (Äthylacetat), des Initiators (O—Cl-Benzoylperoxyd) und des Monomeren bei der Polymerisation von Methylmethacrylat. $T = 60°$ C. Messungen von Palit u. Mitarb. (*101*), ausgewertet nach (V,15); dargestellt nach (V,19)

Die in Abb. 23 an einem Beispiel demonstrierte Bestimmungsmethode der Übertragungskonstanten von Monomerem, Lösungsmittel und Initiator scheint uns besonders empfehlenswert. Mit wenigen (etwa 9—12) gut gewählten Meßpunkten können alle drei Übertragungskonstanten bestimmt werden. Gleichzeitig wird auf diese Weise die für katalysierte Polymerisation unbedingt notwendige Prüfung auf störende Nebenreaktionen durchgeführt (vgl. Abb. 15).

4. Kettenübertragung durch das Polymere. Die Übertragungskonstante des Polymeren kann, wie bereits bei Styrol (Abschnitt Va, 6) ausführlich gezeigt wurde, in derselben Weise bestimmt werden wie jene anderer Kettenüberträger, wenn man die Verminderung des Polymerisationsgrades mißt, welche durch den Zusatz von sehr niedermolekularem Polymerem (α-Polymeres) hervorgerufen wird.

Im Gegensatz zum Polystyrol wurde jedoch gefunden, daß bei Polymethylmethacrylat die formal nach Gl. V,12 ausgerechnete Übertragungskonstante vom Polymerisationsgrad des α-Polymeren abhängt, und zwar in der Weise, daß mit zunehmendem P_α die Übertragungskonstante niedriger wird. Dieser Effekt ist durch die Anwesenheit einer stark übertragenden Endgruppe zu erklären (*129, 62*).

Anstelle von Gl. V,12 gilt also hier:

$$\Sigma \frac{v_{ü}}{v_w} = \left(\Sigma \frac{v_{ü}}{v_w}\right)_0 + C_{Mitte} \frac{[Poly]}{[M]} + C_{Ende} \frac{[Poly]\, K}{[M]\, P_\alpha}. \qquad \text{(V,20)}$$

C_{Mitte} ist die Übertragungskonstante der im Inneren der Kette liegenden Monomereinheiten, C_{Ende} diejenige der Endgruppen. Die Konzentration der übertragungswirksamen Endgruppen ist gegeben durch $K \cdot [Poly]/P_\alpha$, wobei P_α der Polymerisationsgrad des α-Polymeren ist und K angibt, wieviele Endgruppen pro Molekül übertragungsaktiv sind.

Durch Vergleich von Gl. V,20 mit Gl. V,12 sieht man, daß die formal nach letzterer ausgerechnete „Übertragungskonstante des Polymeren" sich zusammensetzt aus 2 Anteilen, nämlich:

$$C_{Poly} = C_{Mitte} + C_{Ende} \frac{K}{P_\alpha}. \qquad \text{(V,21)}$$

Aus einer großen Anzahl von früher veröffentlichten Meßdaten (*62*) rechneten wir mittels Gl. V,14 und Gl. V,12 die „Übertragungskonstante des Polymeren", C_{Poly}, neu aus. In Abb. 24 sind diese Werte gemäß Gl. V,21 gegen den reziproken Polymerisationsgrad des α-Polymeren aufgetragen.

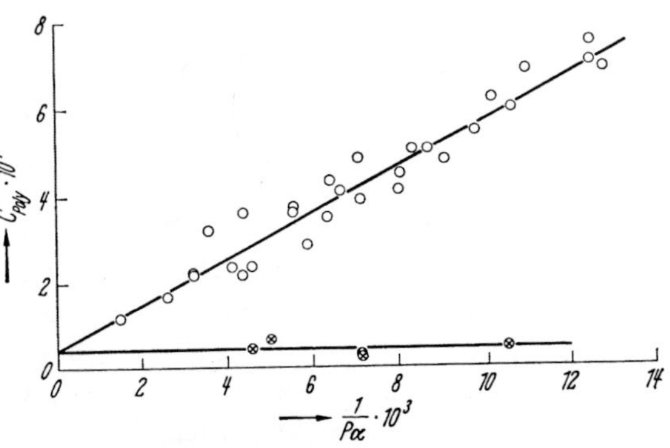

Abb. 24. Kettenübertragung durch das Polymere bei der Polymerisation von Methylmethacrylat. $T = 50°$ C. Eigene Messungen (*62*), dargestellt nach Gl. V,21. ○: radikalisch polymerisierte α-Polymere; ⊗: ionisch polymerisierte α-Polymere

Die Übertragungskonstante der in die Kette eingebauten Methacrylatmoleküle ist nach (V,21) durch den Ordinatenabschnitt der Geraden in Abb. 24 gegeben:

$$C_{Mitte} \cdot 10^5 = 4{,}0 \pm 1{,}5\,.$$

Gegenüber der Konstanten des freien Monomeren ($C_M = 1{,}5 \cdot 10^{-5}$; siehe Abb. 19) tritt also eine schwache Erhöhung auf, die dadurch zu erklären ist, daß die Bindungsverhältnisse an der Methylengruppe sich ändern (im Monomeren: $=CH_2$; im Polymeren: $-CH_2-$).

Ionisch polymerisierte α-Polymethacrylate geben, unabhängig von ihrem Polymerisationsgrad, den reinen Mittelgruppeneffekt (siehe Abb. 24).

Die Steigung der Geraden in Abb. 24 liefert nach Gl. V,21 zunächst $K \cdot C_{Ende}$. Durch Versuche mit „geregelten" Polymeren und solchen, die einem teilweisen thermischen Abbau unterworfen wurden, konnte wahrscheinlich gemacht werden, daß jene Endgruppe, die beim Disproportionierungsabbruch als H-Donator auftritt, die hohe Übertragung bewirkt[1]. Da gleichzeitig auch Kombinationsabbruch stattfindet, ist somit $K < 0{,}5$. Aus der Steigung der Geraden ($5 \cdot 10^{-2}$) erhält man daher für die Übertragungskonstante der Endgruppe:

$$C_{Ende} > 0{,}1 \, .$$

Nach derselben Methode bestimmten Okamura u. Mitarb. (*98*) und Breitenbach u. Mitarb. (*24*) die Übertragungskonstanten von Endgruppen in α-Polymeren, welche sie mit Hilfe von starken Reglern hergestellt hatten. Okamura polymerisierte α-Polystyrol in Tetrachlorkohlenstoff. Er bestimmte (bei 100° C) $C_{Mitte} = 2 \cdot 10^{-4}$. Für die Endgruppe fand er $C_{Ende} = 4 \cdot 10^{-2}$. Dies ist vergleichbar mit der Konstante von Tetrachlorkohlenstoff selbst (vgl. Abb. 11). Breitenbach untersuchte α-Polymere (Polystyrol und Methacrylat), welche in Tetrabromkohlenstoff polymerisiert waren, und fand die Konstante der Endgruppe ebenfalls in guter Übereinstimmung mit jenen von Modellsubstanzen ($C_{Ende} \cong 1{,}6$ bei 60° C). Die starke Übertragungswirkung der Endgruppe wurde zur Herstellung von Blockpolymeren ausgenutzt. Breitenbach u. Mitarb. (*24*) berücksichtigten die Uneinheitlichkeit des Polymeren durch die in Abschnitt I a erwähnte 2-Stufen-Methode.

c) Beispiele von Kettenübertragung bei der Polymerisation von Vinylacetat

Bei diesem Monomeren spielt die Frage, ob der Kettenabbruch durch Kombination oder Disproportionierung erfolgt, eine völlig untergeordnete Rolle. Alle Übertragungskonstanten liegen mindestens eine Größenordnung höher als bei den beiden bisher betrachteten Monomeren. Das hat zur Folge, daß etwa vorkommender Kombinationsabbruch vernachlässigt werden kann und daß man in allen Fällen[2] den Bruchteil der Moleküle, welche aus *einer* Polymerkette bestehen, gleich 1 setzen kann.

Übertragungskonstanten können daher nach den sehr einfachen, für reinen Disproportionierungsabbruch abgeleiteten Gleichungen (Tab. 3 bis 5, Zeile a) bestimmt werden.

Auch das Problem der Eichung der Viscosität zur Erlangung des Polymerisationsgrades liegt hier besonders einfach. Sofern Polymere verwendet werden, die bei niederem Umsatz gewonnen wurden, haben die Eichpräparate stets dieselbe Molekulargewichtsverteilung wie jene Polymeren, deren Molekulargewichte mit Hilfe der Eichgleichung bestimmt werden sollen (nämlich Normalverteilung mit Kopplungsgrad 1). Man erhält daher, wie bereits in Abschnitt I a ausgeführt, immer den Mittel-

---

[1] Durch die seinerzeit durchgeführten Versuche ist jedoch nicht ausgeschlossen, daß die bei der Kombination entstehende Gruppierung die hohe Übertragungskonstante besitzt.

[2] Lediglich bei tieferen Temperaturen könnte der Kombinationsabbruch stärker ins Gewicht fallen. Über sein Ausmaß ist jedoch nichts bekannt.

wert der Eichmethode. Eichung mit Lichtzerstreuung ($\overline{P}_w$) oder Ultrazentrifuge ($\overline{P}_\eta$) ist jedoch aus den bereits mehrfach erwähnten Gründen derjenigen mit der Osmose vorzuziehen. Es sei jedoch nochmals darauf hingewiesen, daß Eichung mit *Fraktionen* in jedem Fall $\overline{P}_\eta$ liefert.

Bei der Polymerisation von Vinylacetat treten nun aber einige Schwierigkeiten auf, die bei den beiden bisher betrachteten Monomeren nicht vorhanden sind.

Die Kinetik der Polymerisation dieses Monomeren ist ganz besonders empfindlich auf Verunreinigungen. Widersprechende Ergebnisse verschiedener Autoren sind sicher häufig auf unterschiedliche Reinigungsoperationen zurückzuführen.

So beobachteten beispielsweise MATSUMOTO u. Mitarb. (*79*) bei Polymerisation in Substanz mit [AIBN] $= 5{,}4 \cdot 10^{-3}$ Mol/l eine Bruttogeschwindigkeit von $1{,}1 \cdot 10^{-3}$ Mol/l sec; hingegen fanden CHADHA und MISRA (*35*) bei derselben Temperatur (60° C) mit der doppelten Konzentration von AIBN praktisch dieselbe Geschwindigkeit. Bei der Polymerisation in Äthylacetat ([AIBN] $\cong 2{,}5 \cdot 10^{-3}$ Mol/l) fand MATSUMOTO (*79*) bei $[M] = 7{,}12$ Mol/l eine Bruttogeschwindigkeit von $4{,}6 \cdot 10^{-4}$; hingegen PALIT und DAS (*100*) bei $[M] \cong 5$ Mol/l $v_{Br} = 4{,}5 \cdot 10^{-5}$.

Obwohl die Messung des UV-Spektrums im Bereich von 2300—2800 Å ein gutes und leicht bestimmbares Maß für die Reinheit des Monomeren bildet (*75*), wird diese Prüfung nur von wenigen Autoren durchgeführt bzw. angegeben. Überdies weisen BENGOUGH und MELVILLE (*14*) darauf hin, daß die UV Absorptionsspektren, die von verschiedenen Autoren bei gereinigtem Vinylacetat beobachtet wurden, nicht übereinstimmen.

Besonders störend wirkt sich auch die Anwesenheit von Sauerstoff bei der Polymerisation von Vinylacetat aus. Während bei Styrol und Methylmethacrylat Bruttogeschwindigkeit und Polymerisationsgrad durch Sauerstoff nicht meßbar beeinflußt werden (*128*, *55*), trifft dies bei Vinylacetat nur für die Bruttogeschwindigkeit zu. Wie von STEIN und SCHULZ (*139*) gezeigt wurde, bilden sich während der Inhibitionsperiode stark übertragungswirksame Substanzen, die den Polymerisationsgrad herabsetzen.

Ferner tritt ein Fehler auf, dessen Einfluß wegen mangelnder Angaben meist nicht richtig abzuschätzen ist. Durch die große Reaktionsgeschwindigkeit bei relativ hoher Polymerisationswärme [20 kcal/Mol (*13*)] kann so viel Wärme freigesetzt werden, daß in Polymerisationsgefäßen üblicher Bauweise (Zylinder von 1—2 cm Durchmesser) eine merkliche Temperaturerhöhung auftritt. Diese beträgt z. B. bei 0,1 Gewichts-% AIBN in Substanz bei 60° C ungefähr 1—2° C (*138*). Es können dadurch erhebliche Fehler auftreten, besonders da diese einen systematischen Gang mit der Monomerkonzentration zeigen (*79*).

Außerdem ist bei Vinylacetat ganz besonders das Problem der Nebenreaktionen zu beachten. Bei Substanzen, welche C–C-Doppelbindungen enthalten, wurde von einer Reihe von Autoren (*104*, *78*, *32*, *30*, *53*, *38*) beobachtet, daß sie außer dem Polymerisationsgrad auch die Bruttogeschwindigkeit herabsetzen. Am Beispiel des Benzols wurde von Stockmayer u. Mitarb. (*141, 104*) sowie von Matsumoto u. Mitarb. (*78*) gezeigt, daß sich dies am besten durch Copolymerisation des Benzols erklären läßt. Auch die von Burnett und Loan (*30*) in dem System [Benzol]/[VAc] = 1 beobachtete Bruttoaktivierungsenergie der Geschwindigkeit von 37,6 kcal/Mol (im Vergleich zu etwa 20 kcal/Mol bei der Polymerisation in Substanz) deutet auf einen völlig anders gearteten Reaktionsschritt hin. Es ist anzunehmen, daß bei Substanzen mit C–C-Doppelbindung nicht normale Kettenübertragung, sondern die Anlagerung eines Radikals an die Doppelbindung der entscheidende Eingriff in den Mechanismus ist.

Aber auch bei Substanzen ohne C–C-Doppelbindung (z. B. Carbonsäureestern) fanden Matsumoto u. Mitarb. (*79*) erhebliche Verzögerungen der Bruttogeschwindigkeit. In verschiedenen Fällen konnten die Autoren nachweisen, daß es sich hierbei um eine Verminderung der Startgeschwindigkeit handelt.

Nach dem Gesagten erscheint es zur Bestimmung von Übertragungskonstanten bei Vinylacetat unerläßlich, durch geeignete Maßnahmen nachzuprüfen, ob es sich in dem betrachteten Fall wirklich um reine Kettenübertragung handelt. Das in Abschnitt Vb, 3 (Abb. 23) an dem System Methacrylat, Äthylacetat, o-Cl-Benzoylperoxyd aufgezeigte Verfahren erscheint hierzu geeignet.

1. Übertragung durch das Monomere. Aus Meßdaten der thermischen Polymerisation kann die Übertragungskonstante des Monomeren bei Vinylacetat nicht berechnet werden. Dixon und Lewis (*42*) sowie Cuthbertson, Gee und Rideal (*40*) berichten, daß völlig aldehydfreies Vinylacetat keine thermische Polymerisation zeigt.

Stockmayer u. Mitarb. (*104*) haben C_M bei 60° aus der katalysierten Substanzpolymerisation bestimmt. Die Autoren verwendeten die Gleichung

$$\frac{1}{\overline{P}_n} = \text{const}\,[I]^{1/2} + C_M\,, \tag{V,22}$$

(Auftragung von $1/\overline{P}_n$ gegen $[I]^{1/2}$), wobei $\overline{P}_n$ aus dem experimentellen $\overline{P}_\eta$ unter Berücksichtigung des Exponenten in Gl. IV,19a berechnet wurde. Gl. V,22 ist identisch mit der sich aus Tab. 3, Zeile a, ergebenden Gleichung:

$$\frac{1}{\overline{P}_n} = \frac{k_{ab}\, v_{Br}}{k_w^2\,[M]^2} + C_M\,, \tag{V,23}$$

unter der Voraussetzung, daß die Initiatorefficiency konstant ist und $[I]$ so niedrig gehalten wird, daß die Änderung in $[M]$ vernachlässigt werden kann. v_{Br} ist dann $\sim [I]^{1/2}$. Das Ergebnis von Stockmayer ist $C_M = 2{,}4 \cdot 10^{-4}$ bei 60° C.

Aus Messungen der Substanzpolymerisation bei verschiedenen Sauerstoffkonzentrationen von STEIN und SCHULZ (*139*) läßt sich der Polymerisationsgrad bei Abwesenheit des Sauerstoffs extrapolieren. Dieser ist $\overline{P}_\eta = 8350$. Mit Hilfe der ebenfalls angegebenen Geschwindigkeit ($2{,}05 \cdot 10^{-4}$ Mol/l sec) und $k_w^2/k_{ab} = 0{,}385$ (vgl. Gl. IV,14) berechnen wir C_M nach der aus Tab. 5, Zeile a, folgenden Gleichung ($a = 0{,}62$; siehe Gl. V,19):

$$\frac{1{,}81}{\overline{P}} - \frac{k_{ab}\, v_{Br}}{k_w^2\, [M]^2} = C_M\,. \qquad \text{(V,24)}$$

Das Ergebnis ist $C_M = 2{,}1 \cdot 10^{-4}$, in guter Übereinstimmung mit dem von STOCKMAYER gefundenen Wert.

Ferner ergibt sich aus den Messungen von MATSUMOTO (*79*) in Lösung (vgl. den nächsten Abschnitt) $C_M \cong 2 \cdot 10^{-4}$, ebenfalls bei 60° C.

AUTRATA und MÜLLER (*4*) verwenden, wie bereits früher (Abschnitt IV b) bemerkt, eine Eichgleichung, welche das Viscositätsmittel des Polymerisationsgrades liefert, betrachten dieses jedoch als Zahlenmittel. Bei Berücksichtigung dieser Tatsache ergibt sich aus den Messungen dieser Autoren bei 60° $C_M \cong$ $\cong 3{,}5 \cdot 10^{-4}$, was im Verhältnis zu den bereits erwähnten Werten etwas hoch erscheint.

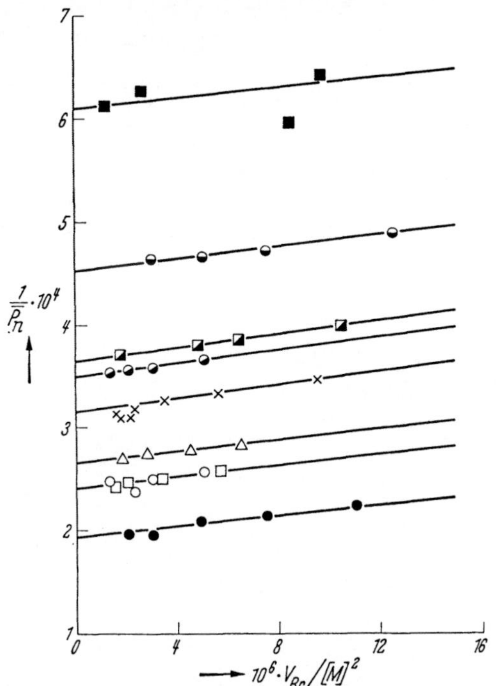

Abb. 25. Polymerisation von Vinylacetat in verschiedenen Estern. Messungen von MATSUMOTO u. Mitarb. (*79*). $T = 60°$ C. Auftragung nach Gl. V,25. ●: Substanz; ×: Äthylacetat [M] = 70%; ○: Isopropylacetat [M] = 85%; ◑: Isopropylacetat [M] = 70%; ⊖: Isopropylacetat [M] = 50%; □: Methylacetat [M] = 85%; ◪: Methylacetat [M] = 50%; ■: Methylacetat [M] = 30%; △: Dimethyloxalat [M] = 70 %. (Mit freundlicher Genehmigung der „Makromolekularen Chemie")

2. Übertragung durch Lösungsmittel. Verschiedene Carbonsäureester und Aldehyde wurden von MATSUMOTO u. Mitarb. (*79*) untersucht. Erhebliche Verzögerung, welche dabei beobachtet wurde, konnte bei einigen der Lösungsmittel auf eine Verminderung der Startgeschwindigkeit zurückgeführt werden. Dies geschah durch Anwendung der Gleichung

$$\frac{2}{\overline{P}_w} = \frac{1}{\overline{P}_n} = \frac{k_{ab}\, v_{Br}}{k_w^2\, [M]^2} + C_M + C_{LM} \frac{[LM]}{[M]} \qquad \text{(V,25)}$$

(vgl. Tab. 3 und 4, Zeile a). Für verschiedene, konstant gehaltene $[LM]/[M]$ ergaben sich bei diesen Lösungsmitteln parallele Geraden bei Auftragung von $1/\overline{P}_n$ gegen $v_{Br}/[M]^2$ (siehe Abb. 25). Dadurch wurde gezeigt, daß keine zusätzliche Abbruchsreaktion die Verzögerung hervorgerufen hatte (vgl. Abschnitt IIIe).

In Abb. 26 ist $\Sigma(v_{ü}/v_w)$, berechnet aus den Meßdaten von MATSUMOTO nach Tab. 3, Zeile a, dargestellt als Funktion von $[LM]/[M]$ für drei verschiedene Ester. Da Messungen mit verschiedener Initiatorkonzentration jeweils auf derselben Geraden liegen, ist auch keine Störung durch eine Nebenreaktion des Initiators (AIBN) vorhanden (vgl. Abb. 15).

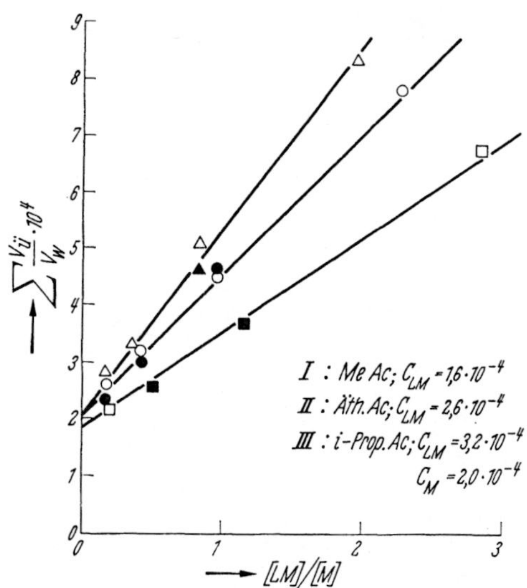

Abb. 26. Übertragung verschiedener Acetate. Messungen von MATSUMOTO (*79*), dargestellt nach Gl. V,5. Methylacetat: [AIBN] = $5 \cdot 10^{-4}$ □; [AIBN] = $10{,}8 \cdot 10^{-4}$ ■; Äthylacetat: [AIBN] = $2{,}65 \cdot 10^{-4}$ ○; [AIBN] = $4{,}85 \cdot 10^{-4}$ ●; *i*-Propylacetat: [AIBN] = $10{,}76 \cdot 10^{-4}$ △; [AIBN] = $2{,}3 \cdot 10^{-4}$ ▲

BUSELLI und LINDEMANN (*32*) untersuchten eine Reihe von Fettsäureäthylestern und fanden, daß die Übertragungskonstanten regelmäßig mit der Zahl n der Kohlenstoffatome in der Fettsäure ansteigen nach der Gleichung: $C_{LM} = 4 \cdot 10^{-3} + n \cdot 0{,}7 \cdot 10^{-3}$. Danach wäre der aliphatischen CH_2-Gruppe eine „partielle Übertragungskonstante" von $0{,}7 \cdot 10^{-3}$ zuzuschreiben. Die gemessenen Konstanten liegen jedoch, im Vergleich zu denjenigen von MATSUMOTO, sehr hoch (z. B. Äthylacetat, 50° C: $C_{LM} = 1{,}2 \cdot 10^{-3}$).

Eine große Anzahl von Lösungsmitteln wurde von PALIT und DAS (*100*) bearbeitet. Die Auswertung erfolgte nach Gl. I,2, wobei $[I]/[M]$ konstant gehalten wurde. Etwaige Veränderungen der Bruttogeschwindigkeit wurden nicht berücksichtigt. Wie bereits erwähnt, haben jedoch mehrere Autoren, besonders in aromatischen Lösungsmitteln, häufig erhebliche Verzögerungen gefunden, die auf Copolymerisation zurückzuführen sind. Aber auch in Bromessigsäure fanden VALE und ROBERTSON (*148*) eine starke Verminderung der Geschwindigkeit.

Ferner wurden die Polymerisationsgrade von PALIT und DAS aus der bei 35° in Aceton gemessenen Viscositätszahl ausgerechnet nach der Gleichung: $[\eta] = 1{,}88 \cdot 10^{-4} M^{0{,}69}$. Diese Beziehung wurde jedoch von WAGNER (*149*) mit Hilfe von Fraktionen der Uneinheitlichkeit $< 0{,}3$ aufgestellt, wobei $[\eta]$ bei 25° in Aceton gemessen wurde. Die hiernach berechneten Polymerisationsgrade sind daher eher Viscositäts- als Zahlenmittel. Das würde bedeuten, daß die von PALIT berechneten Übertragungskonstanten etwa um einen Faktor 1,8 zu niedrig liegen. Der Fehler wird zwar teilweise durch die verschiedene Meßtemperatur kompensiert, jedoch lassen sich exakte $\overline{P}_\eta$ oder $\overline{P}_n$ danach nicht bestimmen.

Entsprechendes gilt auch für Messungen von KAPUR und JOSHI (*68*), die ebenfalls nach (I,2) auswerteten. Als Eichgleichung verwendeten diese Autoren eine Beziehung, die von MCINTOSH u. a. (*85*) mit Hilfe von kommerziellen Polyvinylacetaten verschiedener Uneinheitlichkeit aufgestellt wurde [siehe auch (*149*)]. Die Ergebnisse der beiden Autorengruppen sind nicht in Übereinstimmung. So sind z. B. die Konstanten für Alkohole und Aldehyde bei PALIT (60° C) durchschnittlich um einen Faktor 3—4 größer als bei KAPUR (75° C).

3. Übertragung durch das Polymere. Wie alle anderen Übertragungsreaktionen, ist auch jene am Polymeren selbst stärker ausgeprägt als bei den bisher betrachteten Monomeren. Das verhältnismäßig große Interesse, das gerade dieser Reaktion entgegengebracht wurde, ist vermutlich auf das Bestreben zurückzuführen, stark verzweigte Polymere mit definiertem Verzweigungsgrad in die Hand zu bekommen.

Leider ist bis jetzt keine sehr befriedigende Übereinstimmung zwischen den Ergebnissen verschiedener Autorengruppen zu beobachten. Das liegt vielleicht zum Teil daran, daß bei Polyvinylacetat zwei verschiedenartige übertragungsaktive Stellen vorhanden sind, einmal die Methylgruppen des Acetats, zum anderen die tertiären H-Atome längs der Kette, und daß die verschiedenen verwendeten Untersuchungsmethoden teils auf beide, teils nur auf eine der beiden Verzweigungsarten ansprechen.

WHEELER, LAVIN und CROZIER (*152*) untersuchten die Verzweigung bei Polyvinylacetat mit Hilfe der kinetischen Analyse der Polymerisation bei höheren Umsätzen. Aus ihren Ergebnissen bei 70° C (Tab. 4 der zitierten Arbeit) lassen sich die Konstanten der beiden Übertragungsprozesse abschätzen. Man erhält für die Übertragung längs der Kette $C_{Polv,K} \approx 2 \cdot 10^{-4}$, für die Übertragung an der Acetat-Seitenkette $C_{Polv,S} \approx 4 \cdot 10^{-4}$.

BEVINGTON, GUZMÁN und MELVILLE (*17*) finden bei Verwendung der bereits bei Styrol zitierten, radioaktiven Methode $C_{Polv}(40°) = 3{,}1 \cdot 10^{-3}$; $C_{Polv}(50°) = 3 \cdot 10^{-4}$.

STOCKMAYER u. Mitarb. (*140*) geben $C_{Polv}(60°) = 8 \cdot 10^{-4}$.

PATAT und PATCHINKOV (*103*) bestimmen die „Esterbrücken-Verzweigung" mit Hilfe des Verseifungsabbaues. Aus ihren Meßdaten läßt sich die Übertragungskonstante der Acetat-Seitenkette abschätzen [$C_{Polv,S}(11°) \approx 2 \cdot 10^{-4}$; $C_{Polv,S}(21°) \approx 4 \cdot 10^{-4}$; $C_{Polv,S}(31°) \approx 16 \cdot 10^{-4}$].

AUTRATA und MÜLLER (*4*) schließlich verwenden die Methode der „α-Polymeren", welche bei Styrol und Methacrylat bereits ausführlich beschrieben wurde. Bei Polyvinylacetat liefert sie die Summe der beiden Übertragungskonstanten, $C_{Polv,S} + C_{Polv,K}$. Unter Berücksichtigung der

Uneinheitlichkeit und des von den Autoren verwendeten Konzentrationsmaßes (vgl. das zu Gl. IV,14 Gesagte) kann man aus den Meßdaten $C_{Poly} \approx 2 \cdot 10^{-3}$ (für 50°) ausrechnen. Da jedoch nur ein α-Polymeres untersucht wurde, und somit die Überprüfung der Abhängigkeit der Konstante von $1/P_\alpha$ unterblieb, ist das Vorhandensein eines Endgruppeneffektes nicht auszuschließen (vgl. Abschnitt V b, 4).

d) Tabellarische Zusammenfassung von Übertragungskonstanten

Die folgenden, in Tabellen zusammengefaßten Übertragungskonstanten für Polystyrol (PSty), Polymethylmethacrylat (PMM) und Polyvinylacetat (PVAc) sind nach Stoffgruppen geordnet. Die wiedergegebenen Konstanten sind teilweise den vorangegangenen Abschnitten entnommen. In der Spalte „Zitat" ist dann jeweils die Seite angegeben, auf welcher die ausführliche Berechnung zu finden ist. Bei den übrigen Konstanten ist das Literaturzitat vermerkt.

Ein Strich nach der Nummer des Zitates bedeutet, daß eine Neuberechnung vorgenommen wurde.

Relativ hohe Übertragungskonstanten wurden z. T. auch dann mit in die Tabellen aufgenommen, wenn eine Überprüfung bzw. Korrektur aus Mangel an Meßdaten nicht möglich war. (Wie aus den vorangegangenen Abschnitten hervorgeht, wirken sich etwaige Nebenreaktionen bei starken Übertragungseffekten weniger aus.) Bei solchen Werten sind die Nummern der Literaturzitate mit einem Stern versehen.

Bei einigen Stoffgruppen wurden nur einige typische Vertreter aufgenommen. So haben z. B. Dinaburg und Vansheidt (*41*) und Pierson u. a. (*107*) insgesamt etwa 60 Mercaptane und Disulfide gemessen, von welchen hier nur an einigen wenigen der Einfluß verschiedenartiger Substituenten gezeigt wird.

Tabelle 13. *Übertragungskonstanten von Kohlenwasserstoffen*

Nr.	Überträger	T	$C_X \cdot 10^5$					
		°C	PSty	Zitat	PMM	Zitat	PVAc	Zitat
1	Cyclohexan	60	0,4	S. 539			70	100*
2	Cyclohexan	80	0,8	S. 539				
3	Benzol	80	0,8	S. 539	2,4	S. 552		
4	Toluol	60	1,6	S. 539	4,0	111		
5	Toluol	80	2,7	S. 539	5,2	S. 552		
6	Äthylbenzol	80	11,5	S. 539				
7	Isopropylbenzol	80	18	S. 539				
8	t-Butylbenzol	80	0,8	S. 539				
9	Diphenylmethan	60	28	S. 541				
10	Triphenylmethan	60	65	S. 541				
11	Pentaphenyläthan	60	200000	S. 541				

Zu Tab. 13: Bei PVAc wurden die aromatischen Verbindungen wegen der erwiesenen Copolymerisation (*141*, *104*) und der dadurch entstandenen Unsicherheit der Übertragungskonstanten weggelassen. Auch bei den beiden anderen Monomeren wird sich möglicherweise eine Korrektur als notwendig erweisen, sobald der Copolymerisationseffekt genauer untersucht ist (vgl. Abschnitt Va, 2). Der regelmäßige Anstieg der Konstanten von Toluol (prim. H) über Äthylbenzol (sek. H) zu Isopropylbenzol (tert. H) zeigt jedoch, daß ein wirklicher Übertragungseffekt vorhanden ist. Vergleich von 5, 9, 10 und 11 zeigt den lockernden Einfluß von Phenylgruppen auf das zu übertragende Wasserstoffatom.

Zu Tab. 14: Die von Gregg und Mayo (*51*) bestimmten Konstanten wurden bei der thermischen Polymerisation von Styrol gemessen, wobei die Bruttogeschwindigkeit wenig oder gar nicht von der zweiten Ordnung bzgl. des Monomeren abwich. Wie aus Abschnitt V a, 2 ersichtlich ist, liefert die Auswertung nach der „Mayo-Gleichung" in diesen Fällen sehr nahe dieselben Werte, wie diejenige nach den in der vorliegenden Arbeit verwendeten Gleichungen.

Die Messungen von Palit und Das (*100*) bei PVAc bedürfen einiger Korrekturen (vgl. Abschnitt V c, 2). Sie wurden jedoch mit aufgenommen, um den 1—2 Größenordnungen betragenden Unterschied zu den Konstanten bei PSty und PMM zu veranschaulichen.

Der Vergleich von 2 und 3 zeigt den Unterschied zwischen sekundärem und tertiärem Wasserstoffatom. Interessant ist, daß Morton u. Mitarb. (*94*) durch

Tabelle 14. *Übertragungskonstanten von Alkoholen, Äthern, Säuren und ihren Abkömmlingen sowie Carbonylverbindungen*

Nr.	Überträger	T	$C_x \cdot 10^5$					
		°C	PSty	Zitat	PMM	Zitat	PVAc	Zitat
1	Methanol	60					23	*78a*
2	n-Butanol	60	0,6	*51**			~ 200	*100**
3	i-Propanol	60	8	*94*				
4	t-Butanol	60	2,1	*94*			~ 5	*100**
5	Dioxan	60	0	*51**				
6	Dioxan	90	2	S. 543				
7	Essigsäure	60					~ 10	*100**
8	Propionsäure	60	0,5	*51**				
9	Phenylessigsäure . .	60	60	*51**				
10	Methylacetat	60					16	S. 560
11	Äthylacetat	60			4,6	S. 554	26	S. 560
12	Propylacetat	60					34	*79*
13	i-Propylacetat . . .	60					31	S. 560
14	Methyl-n-butyrat . .	50			$< 10^{-5}$	*61*		
15	Methyl-n-butyrat . .	60					180	*79*
16	Methyl-i-butyrat . .	50			~ 4	*61*		
17	Dimethyloxalat . . .	60					20	*79*
18	Diäthylmalonat . .	60	4,7	*51**				
19	2-Methyl-propionitril	60	18	*94*				
20	Acetaldehyd	60					2000	*79*
21	Butyraldehyd	60	57	*51**			6500	*79*
22	Benzaldehyd	60					2300	*79*
23	Benzaldehyd	104	26	*108*				
24	Aceton	60	< 5	*51**			~ 120	*100**
25	Diäthylketon	60					~ 100	*100**
26	Diäthylketon	80			27	S. 553		

Untersuchungen an deuteriertem t-Butanol nachweisen konnten, daß bei tertiären Alkoholen der Übertragungseffekt durch den Hydroxylwasserstoff hervorgerufen wird.

Bei 8 und 9 ist der lockernde Einfluß der Phenylgruppe auf den α-ständigen Wasserstoff zu beobachten. Carboxyl- und Carbonylgruppen haben offenbar wenig Einfluß (vgl. 2 und 18, sowie 2 und 25). Die sehr ähnlichen Konstanten von 20 und 22 deuten darauf hin, daß bei Aldehyden der Haupteffekt vom Wasserstoffatom der Aldehydgruppe, und nicht vom α-ständigen Wasserstoff hervorgerufen wird.

Zu Tab. 15: Die Überträgerwirkung der Halogene nimmt zu in der Reihenfolge Cl $<$ Br $<$ J entsprechend ihrer Polarisierbarkeit. Dies wird bei Vergleich von 1, 2 und 3 sowie von 13, 14 und 7, 10 anschaulich.

Die Halogenatome von Säurehalogeniden übertragen wesentlich stärker als normale Halogenverbindungen (7, 8).

Halogenierte Benzole zeigen etwa dieselbe Überträgerwirkung wie die entsprechenden halogenfreien Substanzen (vgl. 16, 17, 18 mit Abb. 9).

Über die Wirkung der kernhalogenierten aromatischen Verbindungen ist viel diskutiert worden. BREITENBACH (*20a*) wies zuerst darauf hin, daß in Cl-Benzol hergestelltes Polystyrol kein Halogen enthält. MAYO (*80*) fand bei Brombenzol dasselbe Ergebnis. Zur Erklärung dieses Befundes nimmt MAYO an, daß das Brombenzol ein Wasserstoffatom von der wachsenden Kette auf ein Monomermolekül überträgt, dessen Doppelbindung dadurch geöffnet wird. Kürzlich berichtete BREITENBACH (*21*), daß bei Polyvinylacetat der Polymerisationsgrad durch Brombenzol stärker herabgesetzt wird, als durch Chlorbenzol, wobei die Polymeren jedoch wesentlich weniger Brom als Chlor enthalten. Im Zusammenhang mit den neueren Erfahrungen über die Copolymerisation mit aromatischen Verbindungen (*104*, *37*, *28*) wird vielleicht auch das Problem der Kettenübertragung aromatischer Halogenverbindungen neu untersucht werden müssen.

Zu Tab. 16: Die sehr starke Überträgerwirkung des Mercaptan-Wasserstoffes ist nur wenig beeinflußt von Länge und Art des organischen Restes. Die Nachbar-

Tabelle 15. *Übertragungskonstanten von Halogenverbindungen*

Nr.	Überträger	T	C_X					
		°C	PSty	Zitat	PMM	Zitat	PVAc	Zitat
1	n-Butylchlorid . . .	60	$0{,}4 \cdot 10^{-5}$	*51**				
2	n-Butylbromid . . .	60	$0{,}6 \cdot 10^{-5}$	*51**				
3	n-Butyljodid	60	$18{,}5 \cdot 10^{-5}$	*51**				
4	Dichloräthan	60					$\sim 7 \cdot 10^{-4}$	*100**
5	Dichloräthan	70	$1{,}1 \cdot 10^{-4}$	*25**				
6	Diäthylchlormalonat.	60	$3 \cdot 10^{-3}$	*51**				
7	Acetylbromid	60	0,86	*47**				
8	Bromessigsäure . . .	60	$4{,}3 \cdot 10^{-2}$	*47**				
9	Tribromessigsäure .	90	2,4	*24**				
10	Chloracetylchlorid . .	60	0,33	*47**				
11	Chloroform	60	$5 \cdot 10^{-5}$	*51**			$1{,}3 \cdot 10^{-2}$	*100**
12	Chloroform	80			$1{,}9 \cdot 10^{-4}$	S.553		
13	Tetrachlorkohlenstoff	80	$1{,}7 \cdot 10^{-2}$	S.542	$3{,}3 \cdot 10^{-4}$	S.553		
14	Tetrabromkohlenstoff	60	2,2	*46**	0,3	*46**	> 39	*46**
15	1(Tribrom)-propan .	90	2,4	*24**				
16	m-Dichlorbenzol . .	140	$1{,}4 \cdot 10^{-4}$	*25**				
17	Br-Benzol	155	$3 \cdot 10^{-4}$	*80*				
18	O-Cl-Toluol	70	$6{,}2 \cdot 10^{-5}$	*25**				
19	Chlorwasserstoff. . .	100	0	*81*				

Tabelle 16. *Übertragungskonstanten von Mercaptanen und Disulfiden*

Nr.	Überträger	T	C_X					
		°C	PSty	Zitat	PMM	Zitat	PVAc	Zitat
1	n-Butylmercaptan	60	22	*150*	0,67	*150*	48	*150*
2	n-Butylmercaptan	90	15,4	*41**				
3	n-Hexylmercaptan	90	15,3	*41**				
4	n-Octcacylmercaptan . . .	90	14,7	*41**				
5	n-Dodecylmercaptan . . .	60	18,5	S. 547				
6	i-Propylmercaptan	60			0,38	*97*		
7	t-Butylmercaptan	60	3,6	*48*	0,18	*97*		
8	3-Äthoxypropylmercaptan	60	21	*48*				
9	Äthylmercaptoglykolat . .	60	58	*48*				
10	Butylmercaptoglykolat .	50			0,6	*61*		
11	Äthylmercaptoacetat . . .	60			0,63	*97*		
12	Phenylmercaptan	90	0,08	*41**				
13	Benzylmercaptan	90	25,5	*41**				
14	Bis(2-äthylhexyl)Disulfid .	50	0,005	*107**				
15	Diphenyldisulfid	50	0,06	*107**				
16	Bis(2,6-xylyl)Disulfid . . .	50	0,69	*107**				
17	Schwefelwasserstoff . . .	70	5	*22'*				

schaft eines Phenylringes behindert jedoch die Abspaltung des Wasserstoffatoms (vgl. 12). Hingegen sind bei den Disulfiden die Diarylverbindungen stärker übertragungsaktiv als die Dialkylverbindungen. Wie Tobolsky (*143a*) und Stockmayer (*140a*) durch Verwendung von cyclischen Disulfiden zeigten, handelt es sich hier um die Aufspaltung der S–S-Bindung, vergleichbar der Öffnung der O–O-Bindung bei den Benzoylperoxyden (vgl. Abschnitt V a, 3 und Tab. 18). G. H. Foxley (*45*) machte darauf aufmerksam, daß Diaryldisulfide erhebliche Verminderung der Bruttogeschwindigkeit hervorrufen.

Tabelle 17. *Übertragungskonstanten von Aminen, Phosphinen sowie organometallischen Verbindungen*

Nr.	Überträger	T	C_X					
		°C	PSty	Zitat	PMM	Zitat	PVAc	Zitat
1	Triäthylamin . . .	60	$7{,}1 \cdot 10^{-4}$	*8**	$8{,}3 \cdot 10^{-4}$	*8**	$3{,}7 \cdot 10^{-2}$	*8**
2	Anilin	60	$2 \cdot 10^{-4}$	*51**				
3	Pyridin	60	$6 \cdot 10^{-5}$	*51**				
4	Octylphosphin . .	60	3,6	*105**	2,3	*105**		
5	2-Cyanoäthyl-phosphin	60	5	*105**	1,4	*105**		
6	Bis(2-Cyanoäthyl)-phosphin	60	5	*105**	1,3	*105**		
7	Phenylphosphin . .	60	~ 44	*105**	~ 16	*105**		
8	Dibutylphosphin .	100	2	S.549				
9	Tributylphosphin .	100	$2{,}5 \cdot 10^{-3}$	*106*				
10	Triisobutyl Al . .	100	28,5	*64**				
11	Triäthyl Al	100	17	*64**				
12	Triisobutoxy Al . .	100	$< 10^{-4}$	*64**				
13	Triäthoxy Al . . .	100	$< 10^{-5}$	*64**				

Zu Tab. 17: Ebenso wie bei den Mercaptanen (Tab. 16) ist der Übertragungseffekt des Phosphinwasserstoffes sehr wenig von der Art der organischen Reste beeinflußt (vgl. 4, 5 und 6). Im Gegensatz zum Phenylmercaptan ist Phenylphosphin jedoch ein besonders starker Überträger.

Besonders interessant sind die hohen Konstanten von Trialkyl-Aluminium, die auf die Öffnung von C–Al-Bindungen zurückzuführen sind.

Tabelle 18. *Übertragungskonstanten von Diacylperoxyden (Initiatoren)*

Nr.	Acylgruppe	T	C_X					
		°C	PSty	Zitat	PMM	Zitat	PVAc	Zitat
1	Benzoyl	70	0,075	*39**				
2	Benzoyl	60	0,07	S. 543	0,02	S. 551	0,09	*78a*
3	O-Methyl-benzoyl . .	70	0,175	*39**				
4	O-Fluor-benzoyl . .	70	0,4	*39**				
5	O-Chlor-benzoyl. . .	70	1,91	*39**				
6	O-Chlor-benzoyl. . .	60			0,8	S. 554	0,17	*35**
7	O-Brom-benzoyl . .	70	2,17	*39**				
8	O-Brom-benzoyl . .	60					3,5	*35**
9	m-Brom-benzoyl . .	70	0,46	*39**				
10	m-Brom-benzoyl . .	60					0,6	*35**
11	O-Methyl-benzoyl . .	60			0,06	*111'*		
12	Octanoyl	70	0,1	*39**				
13	Hexadecoyl	70	0,14	*39**				
14	Palmitoyl	60			0,16	*111'*	0,17	*35**
15	Cinnamoyl	70	1,1	*39**				

Zu Tab. 18: Die Übertragungskonstanten der Dibenzoylperoxyde steigen an mit zunehmender Negativität der Substituenten am Benzolkern (vgl. 1, 3, 4, 5, 7). Die Konstanten von aliphatischen Diacylperoxyden sind nicht sehr verschieden von jenen des Benzoylperoxyds (siehe 1, 12 und 13). Konjugierte Doppelbindungen vergrößern jedoch den Effekt (15).

Tabelle 19. *Übertragungskonstanten von Monomerem und Polymerem*

Nr.	Überträger	T	$C_X \cdot 10^5$					
		°C	PSty	Zitat	PMM	Zitat	PVAc	Zitat
1	Monomeres . . .	60	8	S. 536	2	S. 551	22	S. 558
2	Monomeres . . .	50	6	S. 536	1,5	S. 551		
3	PSty	50	20	S. 547	7	*63*		
4	PMM	50	4	*63*	4	S. 554		
5	PMM-Endgruppe	50	10000	*63*	10000	S. 554		
6	PVAc.	50					102	*4**

Zu Tab. 19: In allen drei Fällen ist die Übertragungskonstante des Polymeren höher als jene des Monomeren. Das ist vermutlich auf die Veränderung der Bindungsverhältnisse der Wasserstoffatome beim Übergang: $H_2C{=}CR_2 \rightarrow -CH_2-CR_2-$ zurückzuführen. Erwartungsgemäß ist dieser Unterschied bei PSty und PVAc größer als bei PMM, da bei den beiden ersteren Polymeren tertiäre Wasserstoffatome vorhanden sind.

e) Zur Reaktivität verschiedener Polymerradikale

Wenn man die Übertragungsreaktionen zwischen verschiedenen Polymerradikalen und derselben Überträgersubstanz miteinander vergleichen will, muß man anstelle der „Übertragungskonstante" C_X die Geschwindigkeitskonstante der Übertragungsreaktion, $k_{ü}$, heranziehen. Mit Hilfe der durch die Gleichungen IV,12a, IV,13a und IV,14a gegebenen Werte für die Geschwindigkeitskonstante des Kettenwachstums wird $k_{ü}$ berechnet nach

$$k_{ü} = C_X k_w \,. \qquad \text{(V,26)}$$

In erster Näherung wäre zu vermuten, daß die Konstanten $k_{ü}$ in einem direkten Zusammenhang stehen mit den Reaktivitäten des Polymerradikals und den Bindungsfestigkeiten der zu übertragenden Atome (bzw. Atomgruppen). Man sollte daher erwarten, daß die Geschwindigkeitskonstanten der Reaktionen der drei Radikale PSty*, PMM* und PVAc* mit verschiedenen Überträgersubstanzen etwa in demselben Verhältnis zueinander stehen. Aus Tab. 20 ist jedoch zu ersehen, daß dies nicht der Fall ist.

Tabelle 20. *Verhältnis der Geschwindigkeitskonstanten der Kettenübertragung zwischen den Radikalen PSty*, PMM* und PVAc* und verschiedenen Überträgersubstanzen, bezogen auf* $k_{ü}(PSty^*) = 1$. $T = 60°$ C

Substanz	$k_{ü}$ (PSty*)	:	$k_{ü}$ (PMM*)	:	$k_{ü}$ (PVAc*)
Toluol	1	:	4,5		
Chloroform	1	:	3	:	1400
Triäthylamin	1	:	2,1	:	280
Octylphosphin	1	:	1		
Phenylphosphin	1	:	0,6		
2-Cyanoäthylphosphin	1	:	0,5		
o-Chlor-benzoylperoxyd	1	:	0,8	:	0,5
Benzoylperoxyd	1	:	0,5	:	7
Tetrabromkohlenstoff	1	:	0,25	:	90
t-Butylmercaptan	1	:	0,09		
n-Butylmercaptan	1	:	0,05		
Tetrachlorkohlenstoff[1]	1	:	0,03		

Bei einigen Substanzen ist $k_{ü}$ für PMM etwas höher als für PSty, wie das nach der „allgemeinen Radikalaktivität" [vgl. PRICE (*109*)] zu erwarten ist[2]. Bei anderen Substanzen findet man jedoch $k_{ü}$ (PMM*) sehr viel niedriger als $k_{ü}$ (PSty*).

[1] $T = 80°$ C.

[2] Die Daten für PVAc müssen wegen Unvollständigkeit und Unsicherheit außer acht gelassen werden.

Walling (*150*) hat zuerst darauf hingewiesen, daß bei Übertragungsreaktionen, ebenso wie bei der Copolymerisation, Phänomene der Polarisation eine Rolle spielen [siehe auch Mayo und Walling (*84*), Fuhrmann und Mesrobian (*46*), Bamford und White (*8*) sowie Okamura u. Mitarb. (*69*)]. Man hat sich das so vorzustellen, daß im Übergangszustand ionische Resonanzstrukturen auftreten, die dadurch zustande kommen, daß ein Elektron vom Radikal auf das Überträgermolekül übergeht, bzw. umgekehrt. Diese zusätzlichen Formen leisten einen Beitrag zur Aktivierungsenergie. Polymerradikale mit elektronenanziehenden Substituenten (wie z. B. $-COOCH_3$ bei PMM) werden daher verminderte Reaktionsfähigkeit immer dann aufweisen, wenn die Überträgersubstanz dazu neigt, ionische Resonanzstrukturen unter *Aufnahme* eines Elektrons vom Polymerradikal zu bilden [z. B. Mercaptane (*150*) und Tetrabromkohlenstoff (*46*)]. Eine Erhöhung der Reaktionsfähigkeit des PMM-Radikals hingegen deutet darauf hin, daß zusätzliche Resonanzformen durch Übergang eines Elektrons auf das Radikal zustande kommen [z. B. Triäthylamin (*8*); der Effekt kommt bei Polymerradikalen mit stärker negativen Substituenten, z. B. bei Polyacrylnitril, besser zum Ausdruck, als bei PMM*]

Die Übertragungsreaktion mit *unpolaren* Substanzen (z. B. Kohlenwasserstoffen) sollte danach wirklich ein Maß für die Reaktivität der Polymerradikale darstellen.

Bamford u. Mitarb. (*7*) haben kürzlich diese Überlegungen auf breiter Basis auf die Reaktionen[1] verschiedener Polymerradikale angewendet. Sie kommen zu dem Ergebnis, daß die Geschwindigkeitskonstante k der Reaktion eines Polymerradikals mit einem Substrat gegeben ist durch die Gleichung:

$$\log k = \log k_{ü,T} + \alpha\sigma + \beta .$$

Hierbei ist $k_{ü,T}$ die Geschwindigkeitskonstante der Kettenübertragung zwischen dem betreffenden Radikal und *Toluol* (als Maß für die allgemeine Reaktivität), σ ist die Hammet-Funktion für die Substituenten des Radikals, α und β sind Konstanten für jedes Substrat. Der zweite Term rechts in obiger Gleichung repräsentiert den Einfluß der Substituenten des Radikals auf die Polarisation im Übergangszustand; der letzte Term ist ein Maß für die Festigkeit der zu öffnenden Bindung im Substrat. (Die Konstanten α und β sind in einfacher Weise aus experimentellen Daten bestimmbar und werden von den Autoren für eine Reihe von Substraten gegeben.)

Dieses neue α-β Schema für Radikalreaktionen hat gegenüber dem alten Q-e-Schema den Vorteil, daß es keinen willkürlichen, sondern

[1] Außer der Kettenübertragung wird auch die Copolymerisation berücksichtigt.

einen — zumindest theoretisch — gut definierten Referenzpunkt hat. Ob allerdings die Wahl der Übertragungskonstanten des Toluols als Referenzpunkt für Reaktion ohne polaren Einfluß besonders günstig ist, hängt davon ab, inwieweit die Zuverlässigkeit der Bestimmung von Übertragungskonstanten an aromatischen Verbindungen durch Copolymerisation beeinträchtigt ist. Hierüber ist noch nicht genügend bekannt.

Schlußbemerkung

Das in den Tab. 13—19 zusammengestellte Material an Übertragungskonstanten ist leider noch sehr lückenhaft. Zum Teil liegt das daran, daß einige Arbeiten wegen mangelhafter Angaben der experimentellen Daten nicht berücksichtigt werden konnten. Andererseits macht sich jedoch bemerkbar, daß die Bestimmung von Übertragungskonstanten sich bisher mit wenigen Ausnahmen auf eine bestimmte Gruppe von leicht zugänglichen Substanzen beschränkt hat. Es wäre wünschenswert, daß die Untersuchungen weiter ausgedehnt werden, da die Bestimmung von Übertragungskonstanten wertvolle Hinweise sowohl auf die Reaktivität der Polymerradikale, als auch auf die Festigkeit der bei der Übertragungsreaktion geöffneten Bindung zu geben vermag.

Wir möchten die Gelegenheit benützen, Herrn Prof. Dr. G. V. Schulz für wertvolle Diskussionen und fördernde Kritik herzlich zu danken. Herrn Dr. D. Stein sind wir für weitgehende Mitarbeit bei Abschnitt V c (Polyvinylacetat) verpflichtet. Ferner danken wir Herrn Dr. E. Perry für die Überlassung unveröffentlichter Messungen sowie für Diskussion und verschiedene Literaturhinweise.

Synopsis

The chain transfer reaction in free radical polymerization has been studied in great detail by numerous workers during the past twenty years. One purpose of these investigations was to find suitable substances to regulate the molecular weight of a polymer for practical reasons. Another purpose was to develop information of basic scientific interest. A consideration of chain transfer constants obtained by using different polymer radicals and a variety of chain transfer agents allows one to draw conclusions concerning the reactivity of the radicals as well as the strength of the bond which is broken in the chain transfer step.

A few very important points have been neglected by some authors in their evaluation of chain transfer constants by means of kinetic measurements. Frequently, a retardation of the overall rate is to be observed in the presence of chain transfer agents. A correct value of the chain transfer constant can result only if the reactions which lead to this retardation are properly considered in the kinetic scheme. In addition, the equation which one must use to calculate the chain transfer constant depends on the type of molecular weight average which is measured. Failure to specify or to take these two effects into consideration has made the precise comparison of many chain transfer constants which are reported in the literature of doubtful value. Therefore, a fundamental and comprehensive kinetic treatment of chain transfer reactions is presented, taking into account possible side reactions as well as the points mentioned above.

Equations are developed which allow the calculation of chain transfer constants by using any of the three molecular weight averages which are commonly used ($\overline{P}_n$, $\overline{P}_w$, and $\overline{P}_\eta$). Special consideration is given to those cases in which the mutual destruction of polymer radicals occurs totally or partly by combination (e. g. polystyrene and polymethyl methacrylate). In such polymers, chain transfer causes a broadening of the molecular weight distribution, the degree of broadening being dependent on the amount of chain transfer which occurs.

From a theoretical point of view, the equations which relate the three different molecular weight averages to the chain transfer constant are equally precise. Practically, however, it is recommended that one uses the viscosity average, because it can be measured most accurately.

The equations are summarized in tables at the end of Section III. With the help of these tables, the equation required for the evaluation of the chain transfer constant in any actual case may be selected easily.

In order to demonstrate the application of these equations, examples of the determination of chain transfer constants are given for three cases: (1) polystyrene, in which the mutual destruction of polymer radicals is by combination, (2) polymethyl methacrylate, in which combination as well as disproportionation is important, and (3) polyvinyl acetate, in which chain transfer predominates.

Literature values of chain transfer constants which are reliable according to the standards proposed in this paper are summarized.

Literatur

1. ALLEN, P. W., G. AYREY, F. M. MERRETT and C. G. MOORE: The use of [C^{14}]-labelled initiators in determining the termination reaction in methyl methacrylate free radical polymerization: the importance of molecular weight measurements. J. Polymer Sci. **22**, 549 (1956).
2. — F. M. MERRETT and J. SCANLAN: The interaction of polymerizing systems with rubber and its homologues. Trans. Faraday Soc. **51**, 95 (1955).
3. ANDERSON, D. B., G. M. BURNETT and A. C. GOWAN: Some novel effects in solution polymerization. IUPAC Symposium on Macromolecular Chemistry, Moscow, 1960 Section II 111.
4. AUTRATA, R., u. J. MUELLER: Block- und Pfropf-Copolymere. I. Bestimmung der Konstante der Kettenübertragung durch das Polymere für Vinylacetat. Collection Czechoslov. Chem. Commun. **24**, 3442 (1959).
5. BAGDASARYAN, KH. S., and Z. A. SINITSYNA: Inhibition of polymerization by aromatic compounds. IUPAC Symposium on Macromolecular Chemistry, Moscow 1960, Section II 22.

6. Bamford, C. H., W. G. Barb, A. D. Jenkins and P. F. Onyon: The kinetics of vinyl polymerization by radical mechanism. Butterworth Scientific Publications, London 1958. (Dort weitere Zitate.)
7. — A. D. Jenkins and R. Johnston: Pattern of free radical reactivities. Trans. Faraday Soc. **55**, 418 (1959).
7a. — — — Termination by primary radicals in vinyl polymerization. Trans. Faraday Soc. **55**, 1451 (1959).
7b. — — — and E. F. T. White: The relations between the molecular weight and intrinsic viscosity of polyacrylnitrile. Trans. Faraday Soc. **55**, 168 (1959).
8. — and E. F. T. White: Tertiary amines as chain transfer agents and their use in the synthesis of block copolymers. Trans. Faraday Soc. **52**, 716 (1956).
9. Bartlett, P. D., and H. Kwart: Dilatometric studies of the behavior of some inhibitors and retarders in the polymerization of liquid vinyl acetate. J. Am. Chem. Soc. **72**, 1051 (1950).
10. Basu, S., J. N. Sen and S. R. Palit: Degree of polymerization and chain transfer in methyl methacrylate. Proc. Roy. Soc. **A 202**, 485 (1950).
11. Bawn, C. E. H.: High polymer solutions. Part V. Effect of concentration on the viscosity of dilute solutions. Trans. Faraday Soc. **47**, 97 (1951).
12. Baysal, B., and A. V. Tobolsky: Rates of initiation in vinyl polymerization. J. Polymer Sci. **8**, 529 (1952).
13. Bengough, W. J.: A method of measuring the heats of polymerization. Trans. Faraday Soc. **54**, 54 (1958).
14. — and H. W. Melville: A thermocouple method of following the non-stationary state of chemical reactions. I. The evaluation of velocity coefficients for vinyl acetate, methylmethacrylate and butyl acrylate polymerization reactions. Proc. Roy. Soc. **A 225**, 330 (1954).
15. Benson, S. W., and A. M. North: A simple dilatometric method of determining the rate constants of chain reactions. II. The effect of viscosity on the rate constants of polymerization reaction. J. Am. Chem. Soc. **81**, 1339 (1959).
16. Bevington, J. C.: The sensitized polymerization of styrene. Trans. Faraday Soc. **51**, 1392 (1955).
17. — G. M. Guzmán and H. W. Melville: Self-branching in the polymerization of vinyl acetate. Proc. Roy. Soc. **A 221**, 437 (1954).
18. — — — Self-branching in the polymerization of styrene. Proc. Roy. Soc. **A 221**, 453 (1954).
19. — H. W. Melville and R. P. Taylor: The termination reaction in radical polymerizations, polymerizations of methyl methacrylate and styrene at 25°. J. Polymer Sci. **12**, 449 (1954).
20. Breitenbach, J. W.: Wirkungsmechanismus von Kettenüberträgern und Verzögerern. Z. Elektrochem. **60**, 286 (1956).
20a. — Naturwiss. **29**, 708 (1951).
21. — u. G. Falthansl: Einbau von aromatischen Lösungsmitteln bei der Polymerisation von Vinylacetat. Monatsh. Chem. **91**, 736 (1960).
22. — u. O. F. Olaj: Schwefelwasserstoff als Kettenüberträger bei der Polymerisation. Makromol. Chem. **29**, 139 (1959).
23. — — u. A. Schindler: Zur Kenntnis der thermischen Polymerisationsanregung. Kunststoffe-Plastics **5**, 302 (1958).
24. — — — Darstellung von Blockcopolymeren mittels polyfunktioneller Kettenüberträger. Monatsh. Chem. **91**, 205 (1960).
25. — u. A. Schindler: Zur Kenntnis der Polymerisation des Styrols in Lösung. Monatsh. Chem. **88**, 810 (1957).

26. Breitenbach, J. W. u. W. Schulz: Zur Kenntnis der Einwirkung von Chinonen und Peroxyden auf Styrol. Monatsh. Chem. **80**, 463 (1949).
27. — A. Springer u. E. Abrahamczik: Österr. Chem. Ztg. **41**, 182 (1938).
28. Burnett, G. M.: Gordon Research Conference on Polymers, July 1960, Colby College.
29. — and L. D. Loan: Solvent participation in radical chain reactions. Part 1. Kinetic Analysis. Trans. Faraday Soc. **51**, 214 (1955).
30. — — Solvent participation in radical chain reactions. Part 2. Rates of polymerization in benzene solution. Trans. Faraday Soc. **51**, 219 (1955).
31. — and H. W. Melville: Chain transfer and inhibition. The determination of transfer coefficients in polymerization reactions. Disc. Faraday Soc. **2**, 322 (1947).
32. Buselli, A. J., M. K. Lindemann and Ch. E. Blades: A study of chain transfer in homologous vinyl esters. J. Polymer. Sci. **28**, 485 (1958).
33. Cantow, H.-J., J. Pouyet u. C. Wippler: Die Abhängigkeit der Viscosität von Polymethylmethacrylaten in Lösung vom Geschwindigkeitsgradienten. Makromol. Chem. **14**, 110 (1954).
34. Chadha, R. N., and G. S. Misra: Catalysed polymerization of styrene in toluene. Current Sci. (India) **1954**, 186.
35. — — Peroxides as initiators in the polymerization of vinyl acetate. Trans. Faraday Soc. **54**, 1227 (1958).
36. — J. S. Shukla and G. S. Misra: Studies in chain transfer. Part 2. Catalysed polymerization of methyl methacrylate. Trans. Faraday Soc. **53**, 240 (1957).
37. Chitani, T., Y. Horibe and M. Kobayakawa: Radical polymerization of styrene in heavy p-diethyl benzene. Kobushi Kagaku **15**, 438 (1958).
38. Conix, A., and G. Smets: Benzoyl peroxide initiated polymerization kinetics of vinyl monomers in various solvents. J. Polymer. Sci. **10**, 525 (1953).
39. Cooper, W.: The effect of structure of diacyl peroxides on their radical induced decomposition in vinyl monomers. J. Chem. Soc. **1952**, 2408.
40. Cuthbertson, A. C., G. Gee and E. K. Rideal: The kinetics of polymerization. Nature (Lond.) **140**, 889 (1937).
41. Dinaburg, V. A., and A. A. Vansheidt: Mercaptans and disulfides as chain transfer agents in thermal polymerization of styrene. Zhur. Obschei Khim. **24**, 840 (1954).
42. Dixon, G., and G. Lewis: The polymerization of vinyl acetate. Proc. Roy. Soc. **A 198**, 510 (1949).
43. Elias, H.-G., u. F. Patat: Zur Verzweigung von Polyvinylacetat. Makromol. Chem. **25**, 13 (1958).
44. Flory, P. J.: The mechanism of vinyl polymerizations. J. Am. Chem. Soc. **59**, 241 (1937).
45. Foxley, G. H.: Diaryl disulfides as moderators of the polymerization of styrene. J. Polymer. Sci **41**, 545 (1959).
46. Fuhrmann, N., and R. B. Mesrobian: Chain transfer of vinyl monomers with carbon tetrabromide. J. Am. Chem. Soc. **76**, 3281 (1954).
47. Gannon, Y. A., E. M. Fettes and A. V. Tobolsky: Chain transfer of styrene with various dihalides and the preparation of polystyrene dihalides. J. Am. Chem. Soc. **74**, 1854 (1952).
48. Gregg, R. A., D. M. Aldermann and F. R. Mayo: Chain transfer in the polymerization of styrene. V. Polymerization of styrene in the presence of mercaptans. J. Am. Chem. Soc. **70**, 3740 (1948).

49. GREGG, R. A., and F. R. MAYO: Chain transfer in the polymerization of styrene. III. The reactivities of hydrocarbons toward the styrene radical. Disc. Faraday Soc. **2**, 328 (1947).
50. — — Chain transfer in the polymerization of styrene. II. The reaction of styrene with carbon tetrachloride. J. Am. Chem. Soc. **70**, 2373 (1948).
51. — — Chain transfer in the polymerization of styrene. VII. Compounds containing halogens, oxygen and nitrogen. J. Am. Chem. Soc. **75**, 3530 (1953).
52. HAHN, W., W. MUELLER u. R. V. WEBBER: Untersuchungen an Polystyrolen verschiedener Herstellungen. Makromol. Chem. **21**, 131 (1956).
53. HARDY, G., u. K. NYITRAI: Kinetische Untersuchungen der radikalartigen Polymerisation des Vinylacetats in Gegenwart von fünfgliedrigen heterocyclischen Verbindungen. IUPAC Symposium über Makromoleküle, Wiesbaden 1959, Sektion III B6.
54. HAYDEN, P., and H. MELVILLE: The kinetics of the polymerization of methyl methacrylate. I. The bulk reaction. J. Polymer. Sci. **43**, 201 (1960).
55. HENRICI-OLIVÉ, G., u. S. OLIVÉ: Reaktionskinetik der Polymerisationshemmung durch molekularen Sauerstoff (Versuche mit Styrol). Makromol. Chem. **24**, 64 (1957).
56. — — Kinetische Messungen zum Trommsdorff-Effekt in Polystyrol. Kunststoffe-Plastics **5**, 315 (1958).
57. — — Die Beeinflussung des Polymerisationsablaufs durch Nebenreaktionen der Initiatorradikale I. IUPAC Symposium über Makromoleküle, Wiesbaden 1959, Sektion III B3.
58. — — Die Beeinflussung des Polymerisationsablaufs durch Nebenreaktionen der Initiatorradikale II. Makromol. Chem. **37**, 71 (1960).
59. — — Side reactions in the radical polymerization of styrene. 137th Meeting of the Am. Chem. Soc. Cleveland, Ohio, April 1960, Division of Polymer Chemistry, S. 184.
59a. — — Zur radikalischen Polymerisation von Styrol in Lösungsmitteln verschiedener Dielektrizitätskonstante. Makromol. Chem. **42**, 251 (1960).
60. — — Branching in polystyrene. J. Polymer. Sci. (im Druck).
61. — — Bisher unveröffentlichte Versuche.
62. — — u. G. V. SCHULZ: Übertragungskonstante und Konstitution des Polymethylmethacrylates. Makromol. Chem. **23**, 207 (1957).
63. — — — Selbstverzweigung und Übertragungsreaktion am Polymeren bei Polystyrol. Z. physik. Chem. NF **20**, 176 (1959).
64. HUFF, T., and E. PERRY: The reaction of styryl radicals with organoaluminium compounds. J. Am. Chem. Soc. **82**, 4277 (1960).
65. HUGGINS, M. L.: The viscosity of dilute solutions of long-chain molecules. IV. Dependence on concentration. J. Am. Chem. Soc. **64**, 2716 (1942).
66. JENKINS, A. D.: Transfer to solvent and retardation in vinyl polymerization. Trans. Faraday Soc. **54**, 1885 (1958).
67. KAPUR, S. L.: Chain transfer in the peroxide catalysed polymerization of styrene. J. Polymer Sci. **11**, 399 (1953).
68. — and R. M. JOSHI: Chain transfer in solution polymerization. II. Vinyl Acetate. J. Polymer. Sci. **14**, 489 (1954).
69. KATAGIRI, K., K. UNO and S. OKAMURA: Q and e values for several chain transfer agents. J. Polymer. Sci. **17**, 142 (1955).
70. KICE, J. L.: Inhibition of polymerization. I. Methyl methacrylate. J. Am. Chem. Soc. **76**, 6274 (1954).
71. KIRCHNER, K., u. F. PATAT: Die Startreaktion der unkatalysierten Styrolpolymerisation. Makromol. Chem. **37**, 251 (1960).

72. KUHN, W., u. H. KUHN: Bedeutung beschränkt freier Drehbarkeit für die Viscosität und Strömungsdoppelbrechung von Fadenmoleküllösungen. Helv. Chim. Acta **28**, 1533 (1945).
73. LIM, D., and M. KOLINSKY: Polymerization in solvents of low solvent power. IUPAC Symposium über Makromoleküle, Wiesbaden **1959**, Sektion III B 14.
74. MATHESON, M. S., E. E. AUER, E. B. BEVILACQUA and E. J. HART: Rate constants in free radical polymerization. I. Methyl methacrylate. J. Am. Chem. Soc. **71**, 497 (1949).
75. — — — — Rate constants in free radical polymerizations. II. Vinyl Acetate. J. Am. Chem. Soc. **71**, 2610 (1949).
76. — — — — Rate constants in free radical polymerizations. III. Styrene. J. Am. Chem. Soc. **73**, 1700 (1951).
77. — — — — Rate constants in free radical polymerizations. IV. Methyl Acrylate. J. Am. Chem. Soc. **73**, 5395 (1951).
78. MATSUMOTO, M., and M. MAEDA: Kinetics of the polymerization of vinyl acetate in benzene solution. J. Polymer Sci. **17**, 435 (1955).
78a. — — Chain transfer constant of benzoyl peroxide to polyvinyl acetate radical. J. Polymer Sci. **17**, 438 (1955).
79. — J. UKIDA, G. TAKAYAMA, T. EGUCHI, K. MUKUMOTO, K. IMAI, Y. KAZUSA and M. MAEDA: Effects of esters, aldehydes and some impurities on the polymerization of vinyl acetate. Makromol. Chem. **32**, 13 (1959).
80. MAYO, F. R.: Chain transfer in the polymerization of styrene. VIII. Chain transfer with bromobenzene and mechanism of thermal initiation. J. Am. Chem. Soc. **75**, 6133 (1953).
81. — Free radical addition and transfer reactions of hydrogen chloride with unsaturated compounds. J. Am. Chem. Soc. **76**, 5392 (1954).
82. — The dimerization of styrene. IUPAC Symposium on Macromolecular Chemistry, Moscow 1960, Section II 11.
83. — R. A. GREGG and M. S. MATHESON: Chain transfer in the polymerization of styrene. VI. Chain transfer with styrene and benzoyl peroxide; the efficiency of initiation and the mechanism of chain termination. J. Am. Chem. Soc. **73**, 1691 (1951).
84. — and CH. WALLING: Copolymerization. Chem. Revs. **46**, 191 (1950).
85. MCINTOSH, R., and J. A. MORRISON: The molecular weight of some polyvinyl acetates. Can. J. Research. **B 24**, 192 (1946).
86. MELVILLE, H. W., and L. VALENTINE: Studies in copolymerization. I. The evaluation of the kinetic coefficients of the copolymerization of styrene and methyl methacrylate. Proc. Roy. Soc. **A 200**, 337 (1950).
87. — — Studies in copolymerization. II. Discussion of the validity of the results and some theoretical implications. Proc. Roy. Soc. **A 200**, 358 (1950).
88. MEYERHOFF, G.: Molekulargewichtsbestimmungen an verschieden scharf fraktionierten Polymethacrylsäuremethylestern. Makromol. Chem. **12**, 45 (1954).
89. — Über die Durchschnittswerte der Molekulargewichte makromolekularer Substanzen. Makromol. Chem. **12**, 61 (1954).
90. — Über den experimentellen Zusammenhang zwischen Molekulargewicht und Viskositätszahl von Polystyrolen im Bereich von $M = 1000$ bis 500000 auf Grund von Sedimentations- und Diffusionsmessungen. Z. physik. Chem. NF **4**, 335 (1955).
91. — Zur Bestimmung von Molekulargewichten und ihrer Verteilung bei unfraktionierten Polymeren. Z. Elektrochem. **61**, 1249 (1957).
92. — u. M. CANTOW: Molekulargewichte und Molekulargewichtsverteilungen unverzweigter und verzweigter Polystyrole. J. Polymer Sci. **34**, 503 (1959).

93. MISRA, G. S., and R. N. CHADHA: Studies in chain transfer. Part 3. Catalysed polymerization of styrene. Makromol. Chem. **23**, 134 (1957).
94. MORTON, M., J. A. CALA and J. PIIRMA: The branching reaction. I. Chain transfer of styrene with thiol, alcohol and nitrile. J. Am. Chem. Soc. **78**, 5394 (1956).
95. — and J. PIIRMA: The branching reaction. II. Styrene and methyl methacrylate. J. Am. Chem. Soc. **80**, 5596 (1958).
96. NOZAKI, K., and P.D. BARTLETT: The kinetics of decomposition of benzoyl peroxide in solvents. J. Am. Chem. Soc. **68**, 1686 (1946).
97. O'BRIEN, J. L., and F. GORNICK: Chain transfer in the polymerization of methyl methacrylate. I. Transfer with monomer and thiols. The mechanism of the termination reaction at 60°. J. Am. Chem. Soc. **77**, 4757 (1955).
98. OKAMURA, S., and K. KATAGIRI: Chain transfer to telomer. Makromol. Chem. **28**, 177 (1958).
99. OVERBERGER, C. G., M. T. O'SHAUGHNESSY and H. SHALIT: The preparation of some aliphatic azonitriles and their decomposition in solution. J. Am. Chem. Soc. **71**, 2661 (1949).
100. PALIT, S. R., and S. K. DAS: Studies in chain transfer. IV. Catalysed polymerization of vinyl acetate. Proc. Roy. Soc. **A 226**, 82 (1954).
101. — M. S. NANDI and N. G. SAHA: Studies in chain transfer. III. Determination of chain transfer coefficients from catalysed polymerization data. J. Polymer. Sci. **14**, 295 (1954).
102. PATAT, F.: Zur Kinetik von Reaktionen des Komplextyps. Österr. Chem. Ztg. **61**, 185 (1960).
103. — u. J. A. POTCHINKOV: Zur Verzweigung von Polyvinylacetat I. Makromol. Chem. **23**, 54 (1957).
104. PEEBLES jr., L. M., J. T. CLARKE and W. H. STOCKMAYER: The copolymerization of benzene with vinyl acetate. J. Am. Chem. Soc. **82**, 4780 (1960).
105. PELLON, J.: Phosphines: Chain transfer studies with styrene and methyl methacrylate. J. Polymer Sci. **43**, 537 (1960).
106. PERRY, E.: Private Mitteilung.
107. PIERSON, R. M., A. J. COSTANZA and A. H. WEINSTEIN: Bis-type modifiers in polymerization. I. Behavior of various disulfides in bulk styrene polymerization. J. Polymer Sci. **17**, 221 (1955).
108. PLATAN, G., F. R. EIRICH and R. B. MESROBIAN: Chain transfer of styrene and p-chlorostyrene with benzaldehyde and p-chlorobenzaldehyde. J. Polymer Sci. **39**, 357 (1959).
109. PRICE, CH. C.: Some relative monomer reactivity factors. J. Polymer Sci. **3**, 772 (1948).
110. ROBERTSON, E. E.: Diffusion control in the polymerizations of methyl methacrylate and styrene. Trans. Faraday Soc. **52**, 426 (1956).
111. SAHA, N. G., U. S. NANDI and S. R. PALIT: Peroxides as initiators of polymerization of methyl methacrylate. J. Chem. Soc. **1956**, 427.
112. SANTHAPPA, M., and V. S. VAIDHYANATHAN: Chain transfer reactions in addition polymerization of styrene. Current Sci. (India) **1954**, 259.
113. SCHULZ, G. V.: Über die Beziehung zwischen Reaktionsgeschwindigkeit und Zusammensetzung des Reaktionsproduktes bei Makropolymerisationsvorgängen. Z. physik. Chem. **B 30**, 379 (1935).
114. — Über die Verwendbarkeit des Ostwald-Viscosimeters für die Bestimmung hoher Molekulargewichte. Z. Elektrochem. **43**, 479 (1937).
115. — Über die Kinetik der Kettenpolymerisation. V. Der Einfluß verschiedener Reaktionsarten auf die Polymolekularität. Z. physik. Chem. **B 43**, 25 (1939).

116. Schulz, G. V.: Die Verteilungsfunktionen polymolekularer Stoffe und ihre Ermittlung durch Zerlegung in Fraktionen. Z. physik. Chem. **B 47, 155** (1940).
117. — Zur Klassifizierung der Inhibitor- und Reglerwirkungen bei Polymerisationsreaktionen. Ber. dtsch. chem. Ges. **80**, 232 (1947).
118. — Über den heutigen Stand viskosimetrischer Molekulargewichtsbestimmungen. Kolloid. Z. **115**, 90 (1949).
119. — Die Polymerisation als Testreaktion zur Untersuchung der Kinetik von Radikalreaktionen. Collection Czechoslov. Chem. Commun. **22**, 228 (1957).
120. — Über die Bestimmung des mittleren Molekulargewichts (M_n) von polymolekularen Stoffen. Z. Elektrochem. **60**, 199 (1956).
121. — Über die Polymerisationskinetik in hochkonzentrierten Systemen zur Kinetik des Trommsdorffeffekts an Methylmethacrylat. Z. physik. Chem. NF **8**, 290 (1956).
122. — u. F. Blaschke: Orientierende Versuche zur Polymerisation des Methacrylsäuremethylesters. Z. physik. Chem. **B 50**, 305 (1941).
123. — — Die Polymerisation von Methacrylsäuremethylester unter Einwirkung von Benzoylperoxyd. Z. physik. Chem. **B 51**, 75 (1942).
124. — u. H. J. Cantow: Vorschlag zur Unterscheidung der zwei Größen: „Grenzviskositätszahl" und „konventionelle Viskositätszahl". Makromol. Chem. **13**, 71 (1954).
125. — — u. G. Meyerhoff: Molekulargewichtsbestimmung zu makromolekularen Stoffen. „Methoden der Organischen Chemie" (Houben-Weyl) B. III, S. 375—448.
126. — A. Dinglinger u. E. Husemann: Die thermische Polymerisation von Styrol in verschiedenen Lösungsmitteln. Z. physik. Chem. **B 43**, 385 (1939).
127. — u. J. Harborth: Über den Mechanismus des explosiven Polymerisationsverlaufs des Methacrylsäuremethylesters. Makromol. Chem. **1**, 106 (1947).
128. — u. G. Henrici: Reaktionskinetik der Polymerisationshemmung durch molekularen Sauerstoff (Versuche mit Methylmethacrylat). Makromol. Chem. **18/19**, 437 (1956).
129. — — u. S. Olivé: Die Übertragungskonstante von Polymethylmethacrylat bei der Polymerisation von Methylmethacrylat und Styrol. J. Polymer. Sci. **17**, 45 (1955).
130. — G. Henrici-Olivé u. S. Olivé: Zur quantitativen Berücksichtigung veränderlicher Mittelwerte des Polymerisationsgrades bei der Bestimmung von Übertragungskonstanten. Z. physik. Chem. NF **19**, 125 (1959).
131. — — — Bestimmung der absoluten Geschwindigkeitskonstanten für das Kettenwachstum und die beiden Abbruchsreaktionen, Disproportionierung und Kombination, an Methylmethacrylat. Z. physik. Chem. N. F. **27**, 1 (1961).
132. — u. W. H. Kuhn: Zur Kinetik des Einstellvorganges im Osmometer. Makromol. Chem. **29**, 220 (1959).
133. — u. G. Meyerhoff: Bestimmung der Knäueldimensionen von Fadenmolekülen in Lösungen aus Reibungsdaten. Z. Elektrochem. **56**, 904 (1952).
134. — u. G. Sing: Über den Anstieg der spezifischen Viscosität makromolekularer Lösungen im Bereich kleiner Konzentrationen. J. prakt. Chem. **161**, 161 (1943).
135. Signer, R., u. K. Berneis: Ein Kapillarviskosimeter für kleine Strömungsgradienten. Makromol. Chem. **8**, 268 (1952).
136. Smets, G., and R. Hart: Block and graft copolymers. Fortschr. Hochpolym. Forsch. **2**, 173 (1960).
137. Smith, W. V.: Regulator theory in emulsion polymerization. I. Chain transfer of low molecular weight mercaptans in emulsion and oil phase. J. Am. Chem. Soc. **68**, 2059 (1946).

138. Stein, D.: Private Mitteilung.
139. — u. G. V. Schulz: Übertragungseffekt durch Polyperoxyd bzw. dessen Zerfallsprodukte bei der Polymerisation von Vinylacetat. Makromol. Chem. **38**, 248 (1960).
140. Stockmayer, W. H., J. T. Clarke and R. O. Howard: See P. J. Flory "Principles of polymer chemistry". S. 259. Ithaca N.Y.: Cornell University Press 1953.
140a. — R. O. Howard and J. T. Clarke: Copolymerization of Vinylacetate with a cyclic disulfide. J. Am. Chem. Soc. **75**, 1756 (1953).
141. — and L. H. Peebles jr.: Copolymerization of benzene with vinyl acetate. J. Am. Chem. Soc. **75**, 2278 (1953).
142. Tasset, G., and G. Smets: Degradation of vinyl polymers in solution. J. Polymer Sci. **12**, 517 (1954).
143. Thompson, C. F., W. S. Port and L. P. Witnauer: Chain transfer constants of vinyl esters with toluene. J. Am. Chem. Soc. **81**, 2552 (1959).
143a. Tobolsky, A. V., and B. Baysal: The reaction between styrene and ring disulfides: copolymerization effected by the chain transfer reaction. J. Am. Chem. Soc. **75**, 1757 (1957).
144. Trementozzi, Q. A.: Effect of long-chain branching on some solution properties of polyethylene. J. Polymer Sci. **23**, 887 (1957).
145. Trommsdorff, E., H. Koehle u. P. Lagally: Zur Polymerisation des Methacrylsäuremethylesters. Makromol. Chem. **1**, 169 (1947).
146. Tuedos, F., V. Fuerst u. M. Azori: Inhibition der Styrolpolymerisation mit stabilen freien Radikalen. IUPAC Symposium über Makromoleküle, Wiesbaden 1959, Sektion III B 4.
147. — I. Kende and M. Azori: Kinetics of inhibition of styrene polymerization by nitro compounds. IUPAC Symposium on Macromolecular Chemistry, Moscow 1960, Section II 31.
148. Vale, R. L., and W. G. P. Robertson: Polymerization of vinyl acetate in the presence of bromoacetic acid. J. Polymer. Sci **33**, 519 (1958).
149. Wagner, R. H.: Intrinsic viscosities and molecular weights of polyvinyl acetates. J. Polymer Sci. **2**, 21 (1947).
150. Walling, Ch.: The use of S^{35} in the measurement of transfer constants. J. Am. Chem. Soc. **70**, 2561 (1948).
151. — Free radicals in solution. New York: John Wiley Sons 1957. (Dort weitere Zitate.)
152. Wheeler, O. L., E. Lavin and R. N. Crozier: Branching mechanisms in the polymerization of vinyl acetate. J. Polymer Sci. **9**, 157 (1952).
153. Yoshida, T.: Polymerization of styrene in heavy alcohol and heavy benzene. Bull. Chem. Soc. Japan **23**, 209 (1950).

Fortschr. Hochpolym.-Forsch., Bd. 2, S. 578—595 (1961)

The Anionic Polymerization of Caprolactam

By

O. WICHTERLE, J. ŠEBENDA and J. KRÁLÍČEK

Institute of Macromolecular Chemistry of the Czechoslovak Academy of Science and Dept. of Organic Technology, Technical University, Prague

With 5 Figures

Contents

The base catalysed caprolactam polymerization differs substantially in various aspects from the wellknown industrial polymerization process, which is introduced by water, and the mechanism of which is governed by endstanding carboxylic and amino groups. Therefore it is evident, that the broad experience gathered in the hydrolytic polymerization of caprolactam cannot be applied to the base polymerization, which was developed separately and which requires an independent theoretical treatment.

The strong catalytic activity of bases in the caprolactam polymerization was recognized about as soon (*44—47*) as the "normal" hydrolytic process was, but nevertheless, the fundamental informations about the basic process were disclosed only during the last few years. It was shown quite recently, that the base induced polymerization consists of an extremely complex action of the strong base (*15*, *31*, *66—68*) which is connected with the presence and/or formation of certain additional components. The base plays not only a role in the initiation and propagation reaction but it also involves important side reactions which in turn affect the active centers of the polymerization (*67*, *96*).

The complexity of the base catalysis and its sensitivity towards extremely small quantities of impurities (*25*, *67*) e. g. water, amines, alcohols, caused a lot of mistakes in the previous experimental work and the non-reproducibility was claimed as a characteristic phenomenon of this reaction (*28, 102*). The reaction mechanism was very often interpreted in an incorrect way. The discussion of the process with respect to the technological development met a strict scepticism (*28, 79*) which seemed to be quite understandable under the former poor knowledge of the process.

1. The active bases

Taking into account the slightly acidic nature of the amide group it is to be assumed that caprolactam is converted by addition of strong bases into its salt, e. g. sodium caprolactam according a general scheme

$$\left(\begin{matrix} CO \\ | \\ NH \end{matrix} + \mathbf{NaB} \rightarrow \left[\left(\begin{matrix} CO \\ | \\ N \end{matrix} \right. \right] Na + \mathbf{B{-}H}$$

If any base whose basicity exceeds the basicity of the caprolactam anion is applied, this kind of neutralization reaction results in formation of caprolactam salt, i. e. practically a metal salt of caprolactam. As far as the simultaneously formed component B–H does not interfere to the further catalytic action, especially if this compound is higly volatile and can be easily removed before having opportunity of such interference, the same result is obtained no matter what sort of base was used. Thus, extremely strong bases as alkali metals (*7, 15, 18—22, 31, 42, 43, 46, 47, 49—51, 58, 71, 72, 100, 101*), other strongly elektropositive metals (*42*), metal hydrides (*12, 31, 32, 51, 52*) or metal amides react in a smooth way according the above scheme. The same result i. e. the formation of a caprolactam salt is obtained starting with relatively weak bases as alkali hydroxides (*1, 5, 6, 10, 12, 13, 16—19, 28, 29, 33—35, 37—41, 44, 45, 51, 54, 60, 64, 70—72, 78, 79, 100, 102*) or alcoxides (*9, 45*), leading to an equilibrium

$$\left(\begin{matrix} CO \\ | \\ NH \end{matrix} + NaOH \rightleftharpoons \left[\left(\begin{matrix} CO \\ | \\ N \end{matrix} \right. \right] Na + H_2O$$

in which water or alcohol having about the same acidity as caprolactam has to participate in a similar way in the protonation of the base. Even in this case the equilibrium can be shifted to the right side. Using alkali hydroxides the lactam salt formation is bound to a very quick removal of the water formed, otherwise the simple neutralization is accompanied by a saponification of caprolactam (*62, 70, 102*)

$$\left(\begin{matrix} CO \\ | \\ NH \end{matrix} + NaOH \rightarrow \left(\begin{matrix} COONa \\ \\ NH_2 \end{matrix} \right.$$

This side reaction converts the active strongly basic alkali caprolactam into the aminoacid salt, which is not basic enough to induce the rapid polymerization. Therefore, the addition of alkali hydroxides induces a very fast polymerization only if the water formed is withdrawn as soon as possible either by boiling it out (*35, 64, 79*) or by bubbling a stream of dry nitrogen through the molten lactam (*36, 45*).

There is no analogous ring opening reaction if alkali alcoxides are used and thus they may be successfully used for the preparation of pure salts of caprolactam (*8, 61, 66, 68, 92, 95, 97, 103*) in analogy to the preparation of sodium benzamide (*11, 74, 75*). Although the alcohols formed do not affect directly the salt formed, it is necessary to remove them by distillation in order to avoid side reactions taking place in the polymerization process in the presence of alcohols and esters (see below). In the same way it is possible to prepare alkali salts of other N-alkylamides e. g. benzanilide which also may be used as basic catalysts.

The most simple way of preparing sodium caprolactam consists in dissolving the sodium metal in molten caprolactam. Although the solutions thus obtained are capable to induce a rapid polymerization, we meet with complications arising from the presence of reduction products formed such as hexamethyleneimine (*53, 69, 80, 81*) and 6-aminohexanol (*59, 69*). These by-products affect the course of the subsequent polymerization as they react with the active centers. The ratio neutralization/reduction in the alkali metal reaction with caprolactam depends strongly on the temperature. According to HAMANN (*41*) and YUMOTO (*101*) the reduction reaction takes place to 75% and 10—20% respectively. In order to suppress the reduction it is recommended to maintain the temperature during the dissolution of the sodium metal below 100° (*42*).

In an indirect way the caprolactam salts may be formed by heating caprolactam solutions or suspensions of certain neutral or only slightly basic salts which are able to decompose to strong bases (*82, 83, 86, 87, 91, 93*). This effect can be reached by using alkali salts of carboxylic acids tending to decompose by decarboxylation. Alkali carbonates (*1, 10, 13, 17, 28, 44, 45, 60, 65, 82—84, 87, 91, 102*), alkali salts of carbonic monoalkyl esters, phenylacetic acid and other similar salts (*62, 86, 89, 93, 102*) are converted by decarboxylation into strongly basic carbonium salts or metal derivatives of amines or alcohols

$$C_6H_5\text{–}CH_2\text{–}COONa \rightarrow C_6H_5\text{–}CH_2Na + CO_2$$

$$C_6H_5\text{–}NH\text{–}COONa \rightarrow C_6H_5NHNa + CO_2$$

$$CO\begin{cases} OCH_3 \\ OK \end{cases} \rightleftharpoons CH_3OK + CO_2$$

which are ready to neutralize caprolactam to its alkali salt

$$C_6H_5\text{–}CH_2Na + \underbrace{CO\text{–}NH}_{\smile} \rightarrow C_6H_5\text{–}CH_3 + \left[\underbrace{CO\text{–}N}_{\smile}\right] Na$$

Most of the compounds mentioned are stable at normal temperature, soluble in molten caprolactam and very little sensitive to water. The decarboxylation being a relatively slow process offers the possibility of removing the last traces of moisture from the reaction mixture directly in the reaction vessel at a temperature not high enough to decompose the salt but sufficient to withdraw the remaining water at low pressure (*93*). The slow decomposition of these salts is equivalent to slow metering of a strong base during the polymerization process and this fact allows the polymerization to be carried out by a minimal amount of the strong base (*89*). This might be valuable if high molecular weight polyamides have to be prepared; the strong base, which participates in the end group formation of the polymer (*67*) reduces the final molecular weight (*50*). Therefore, starting the polymerization by sodium phenylacetate or sodium carbonate, the strong bases are formed by decarboxylation very slowly; interrupting the heating just after having reached practically the polymerization equilibrium we can obtain polyamides of the highest degree of polymerization not yet attained by an other method.

The polymerization can be induced even by carboxylic acid salts which are less able to decarboxylate, such as sodium acetate or sodium 6-aminocapronate (*33*, *70*, *71*, *102*); in such case the polymerization starts only at unusually high temperatures, i. e. under conditions which are no more typical to the fast polymerization at relatively low temperatures produced by stronger bases.

Being very strong bases, organometallic compounds are capable to start the polymerization (*1*, *12*, *30*, *51*, *62*). If organomagnesium compounds are used, the interaction of magnesium cations with lactam anions can probably reduce the dissociation of the salt; analogous the basic character is certainly weakened in lactam complexes formed by addition of hydrated oxides of Ti, Zr, Hf, Th or Ce (*48*).

2. The mechanism of the polymerization

In the very first paper by HANFORD and JOYCE (*42*) on sodium polymerization of caprolactam a formation of two end groups of the growing polymer was proposed: (1) a lactamring attached to the open chain by an imidic bond, and (2) an endstanding group –NHNa containing an anion derived from a primary amino group

$$\begin{matrix} \frown CO \\ \,\,| \\ \smile N\text{–}CO \end{matrix} \sim\!\sim\!\sim\!\sim\!\sim NHNa$$

It was assumed, that the latter strongly basic group is responsible for the propagation reaction, being capable to attach a molecule of caprolactam and open its ring structure while regenerating the active basic group (*42*). The assumption that polymerization and degradation processes are caused by –NHNa groups, was accepted by some other authors (*29, 63*); this view is inconsistent with the fact, that in the equilibrium

$$\text{–NHNa} + \text{–CO–NH–} \rightleftharpoons \text{–NH}_2 + [\text{–CO–N–}]\,\text{Na} \qquad \text{(a)}$$

nearly all groups –NHNa must disappear not only because of the enormous prevalence of amidic groups, but above all because the –NH^- anions are incomparably stronger basic than $[\text{–CO–N–}]^-$ anions (*87, 90, 91*). The existence of the salt which is derived from a primary amino group, –NHNa, in an amidic medium is inconsistent in the same way as the existence of sodium alkoxide in glacial acetic acid would be. Moreover, it has been proved, that primary amino groups don't participate in the active polymerization mechanism, because additions of primary amines don't support the anionic polymerization; on the contrary, they cause an appreciable decrease in polymerization rate (*67*).

The presupposition, that carboxylate anions are the active centers of polymer growth (*43, 70, 72*) is equally to be rejected: the addition of carboxylic acid salts, e. g. sodium acetate is without any effect on the polymerization rate (*103*). Besides, the concentration of carboxylic groups as found in the resulting polymers used to be very low (*28, 29, 60*).

The fact, that almost all basic functions in caprolactam or polyamide medium occur in the form of amidic anions $[\text{–CO–N–}]^-$ arose the idea that these particular anions are directly responsible for the polymerization reaction. In analogy to the transesterification mechanism an intermediate addition complex might be formed by combining the amidic group with the respective anion

$$\text{–C}(=\text{O})\text{–OR} + {}^-\text{OR}' \rightleftharpoons \text{–C}(\text{O}^-)(\text{OR})\text{–OR}' \rightleftharpoons \text{–C}(=\text{O})\text{–OR}' + {}^-\text{OR}$$

$$\text{R}_1\text{–C}(=\text{O})\text{–NH–R}_2 + {}^-\text{N}(\text{R}_3)\text{–CO–R}_4 \rightleftharpoons \text{R}_1\text{–C}(\text{O}^-)(\text{N}(\text{R}_3)\text{–CO–R}_4)\text{–NH–R}_2 \rightleftharpoons \text{R}_1\text{–C}(=\text{O})\text{–N}^-\text{–R}_3 + \text{R}_4\text{–CO–NH–R}_2$$

There was a good agreement oft such an interpretation with kinetic measurements of sodium carbonate catalyzed polymerization (*84, 87, 91*). This idea of a simple anionic interchange reaction between an amide group and an amide anion was very attractive, because it allowed to extrapolate the possibility of macroring formation by a simple building-in of caprolactam units into the amide bond. However, basic endgroups (probably amino groups) were found by titrations in every case of base catalyzed polymerization, even in extremely carefully experiments excluding any trace of impurities (*103*). It was necessary to assume, that the formation of end groups is inherent in the polymerization mechanism.

More correct ideas about the polymerization came out after the observation was made, that the polymerization induced by sodium caprolactam starts with a characteristic induction period (*67, 68*). Beginning with a practically zero rate, the polymerization goes on with increasing velocity up to a maximal value (curve A on fig. 1). The induction period indicates that certain active groups are formed by an independent process. The previous assumption of a disproportionation reaction adopted by HANFORD and JOYCE (*42*) can be supported by model reactions like the disproportionation of acetamide, benzamide (*24, 57*) or N-butylacetamide (*68*) in the presence of the respective sodium salts.

$$\begin{array}{ll} C_2H_5{-}N^- & \overset{O}{\overset{\|}{C}}{-}CH_3 \\ \quad | & \quad | \\ CH_3{-}\underset{\underset{O}{\|}}{C} & HN{-}C_2H_5 \end{array} \rightleftharpoons \begin{array}{l} C_2H_5{-}N{-}\overset{O^-}{\overset{|}{C}}{-}CH_3 \\ \quad\;\, | \qquad\; | \\ CH_3{-}\underset{\underset{O}{\|}}{C} \;\; HN{-}C_2H_5 \end{array} \rightleftharpoons \begin{array}{l} C_2H_5{-}N\begin{matrix}\diagup CO{-}CH_3 \\ \diagdown CO{-}CH_3\end{matrix} \\ \quad + \, ^-NH \\ \qquad\;\; | \\ \qquad C_2H_5 \end{array} \qquad \text{(b)}$$

The amide anion formed ($-NH^-$) is immediately neutralized by reaction (a), so that this disproportionation (b) results in the formation of a primary amine and an N-alkylimide of the general structure $R_1\,CO{-}\underset{\underset{R_3}{|}}{N}{-}COR_2$ (e. g. ethyl diacetamide).

Applying this reaction to caprolactam we can imagine, that the first step would be about the same, as HANFORD and JOYCE have proposed. The only correction to be introduced is the above mentioned neutralization of $-NHNa$ to a primary amino group (*87, 90, 91*).

$$\left[\begin{array}{l} N^- \\ | \\ C{=}O \end{array}\right](CH_2)_5 \;\; \left[\begin{array}{l} O{=}C \\ | \\ HN \end{array}\right](CH_2)_5 \rightleftharpoons \left[\begin{array}{l} N{-}\overset{O^-}{\overset{|}{C}} \\ | \qquad | \\ C \quad\; HN \\ \| \\ O \end{array}\right] \; (CH_2)_5 \;\; (CH_2)_5 \rightleftharpoons$$

$$\rightleftharpoons (CH_2)_5\left[\begin{array}{l} N{-}CO{-}(CH_2)_5{-}NH^- \\ | \\ CO \end{array}\right] \xrightarrow{[-CO-NH-]}$$

$$\rightarrow (CH_2)_5\left[\begin{array}{l} N{-}CO{-}(CH_2)_5{-}NH_2 \\ | \\ CO \end{array}\right] + [-CO{-}N{-}]^-$$

It is evident, that by this disproportionation only very low concentrations of disproportionation products (*68*) will result in view of the experience that the opposed reaction, i. e. the reaction of imides with amines takes place very easily (*92*). Assuming this disproportionation as the introductory step in the polymerization mechanism and excluding the participation of the amino group, there remains only the imidic group which might take part in the very fast polymerization reaction (*14, 15, 31, 67*). This possibility was strongly supported by the discovery, that after addition of various N-alkylimides the induction period observed in the base induced polymerization completely disappeared, and that even very small amounts of these imides accelerated the polymerization many times (*66—68*) (curve B on fig. 1). The same effect was found with pyrrolidone (*2, 4, 6, 23, 54, 55*) and piperidone (*6*) polymerization, the mechanism of this reaction being explained in the last years in a similar manner (*14, 15, 31, 100*).

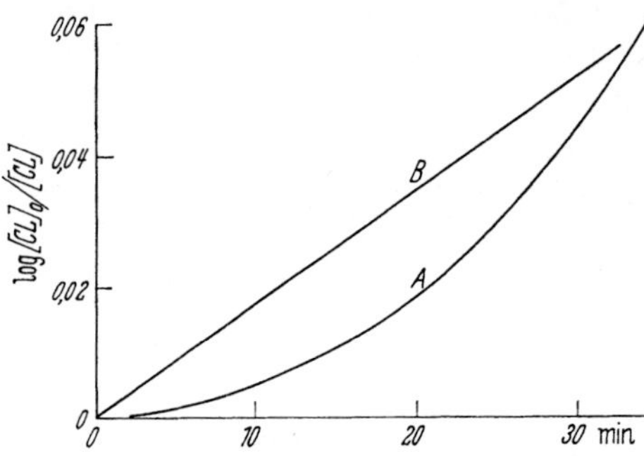

Fig. 1. Polymerization of caprolactam at 192,6°; (*A*) with 0,5 mol-% sodium caprolactam and (*B*) with 0,05 mol-% sodium caprolactam + 0,025 mol-% tetraacetyl hexamethylenediamine

The N-alkylimides alone, without the addition of a strong base have no polymerization activity under similar conditions (low temperature). With respect to the zero polymerization rate in the very beginning of the reaction induced by sodium caprolactam only, we can conclude, that for the typical fast polymerization a simultaneous action of both, imide and alkali caprolactam is essential (*67, 68*).

The mechanism of the caprolactam polymerization, i. e. the transamidation reaction catalysis by the system imide + salt can be interpreted by a nucleophillic attack of the amide anion on the carbonyl group of the imide which represents the strongest electrophillic reagent in the polymerizing system.

A symmetrical intermediate thus formed is subsequently split either into the original parts or into another couple containing interchanged chains. According to this general transamidation scheme (c) we can imagine both growth and depolymerization of a polycaprolactam molecule at its electrophillic imide end closed by an acylated lactam ring (d)

$$\begin{array}{ccc} R_1\;\; R_2\;\; R_4 & & R_1\;\; O^-\;\; R_4 & & R_1\;\; R_2\;\; R_4 \\ \text{N–CO}\;\; {}^-\text{N} & \rightleftharpoons & \text{N—C—N} & \rightleftharpoons & \text{N}^-\;\; \text{CO–N} \\ \text{CO}\qquad \text{CO} & & \text{CO}\;\; R_2\;\; \text{CO} & & \text{CO}\qquad \text{CO} \\ R_3\qquad R_5 & & R_3\qquad R_5 & & R_3\qquad R_5 \end{array} \qquad (c)$$

$$\sim\!\!\sim\!\overset{\overset{O}{\|}}{C}\!-\!\underset{\cup}{N\!-\!\overset{\overset{O}{\|}}{C}}\quad {}^{-}\underset{\cup}{N\!-\!CO} \;\rightleftharpoons\; \sim\!\!\sim\!\overset{\overset{O}{\|}}{C}\!-\!\underset{\cup}{N\!-\!\overset{\overset{O^{-}}{|}}{C}}\!-\!\underset{\cup}{N\!-\!\overset{\overset{O}{\|}}{C}} \;\rightleftharpoons \qquad \text{(d)}$$

$$\rightleftharpoons\; \sim\!\!\sim\!\overset{\overset{O}{\|}}{C}\!-\!\underset{\smile}{N^{-}\;\overset{\overset{O}{\|}}{C}}\!-\!\underset{\cup}{N\!-\!\overset{\overset{O}{\|}}{C}}$$

The cocatalytic activity of N-alkylimides does not depend very much on the nature of residues attached to the imide group. The only striking difference was found with cyclic imides containing both carbonyl groups in a five or six membered ring (*67*), e. g. N-methylsuccinimide or N-methylglutarimide; such substances are no more able to induce a rapid polymerization in the presence of alkali caprolactam. It is abvious, that secondary imides of the general formula R_1–CO–NH–CO–R_2 are ineffective due to their acidity which neutralizes the strong basic lactam salt.

A similar effect as observed with N-alkylimides can be obtained if acylating reagents are added into the lactam to depolymerized (*3*, *9*, *34*, *52*, *53*, *68*), e. g. ketene, acetylchloride; in the presence of sodium caprolactam even carbon monoxide reacts as a formylating agent and therefore a prominent increase in polymerization rate is effected if carbon monoxide is introduced into the alkaline melt of caprolactam (*51*). This kind of cocatalytic activity is probably a general property of derivatives obtained by substitution of the hydrogen atom in N-alkylamides by electron attracting groups (*66*), e. g. by –CO–NH–R or –COONa.

Following the general transamidation reaction between an anion $[-CO-N-]^{-}$ and a carbonyl group from an imide [scheme (c)] a large set of different secondary reactions could proceed (*15*, *67*). We can imagine the formation of amide anions not only from monomeric caprolactam, but anywhere in the growing polymer molecule. Therefore, larger rings might be formed by an intramolecular transamidation reaction.

$$\sim\!CO\!-\!N^{-} \quad \underset{\cup}{CO\!-\!N\!-\!CO}\!\sim \;\rightleftharpoons\; \sim\!CO\!-\!\underset{\underset{CO}{|}}{N}\!\sim \quad \sim\!N^{-}\!-\!CO\!\sim \qquad \text{(e)}$$

Branching of polymer chains might be introduced by intermolecular reactions

CO–N$^-$

CO
N–CO ⇌

CO–N
CO
⇌ N$^-$–CO (f)

It may well be that branched and large ring structures could arise not only from the propagation reaction (scheme (c)), but partly also from the disproportionation reaction (b) which certainly proceeds simultaneously with the propagation. The interaction of a linear amide group with a linear amide anion would generate by disproportionation a branched polymer structure

CO $^-$N
NH CO
[–CO–NH–] →
CO–N
CO
NH$_2$
+ [–CO–N–]$^-$ (g)

The disproportionation reaction takes place with a much lower rate than the propagation reaction and therefore, the rate determining step is the disproportionation if only sodium caprolactam is used as catalyst (*15, 30, 31, 67*). The addition of imides makes it unnecessary to provide the imide by the slow disproportionation reaction and hence the enormous increase of the polymerization rate can be attained by additions of N-alkylimides.

Because the reactions (b) and (c) cannot proceed, if there is no hydrogen at the amide group, N-alkylated caprolactams failed to polymerize by basic catalysts. Indications of N-methyl caprolactam polymerization and copolymerization are to be revised; experiments describing such polymerizations couldn't be reproduced and the authors mentioned seem to be mislead by results obtained by the polymerization of samples of N-methylcaprolactam containing appreciable amounts of unsubstituted caprolactam.

3. Side reactions

The action of bases on caprolactam or amide groups of the polymer molecule is not limited to the disproportionation and transamidation reactions which are directly involved in the polymerization mechanism. A more detailed study of the polymerization kinetics has shown that the catalytic activity of the system caprolactam salt + N-alkylimide expires after a certain time (*67, 92, 96*).

Nonpolymerizable cyclic imides like N-methylsuccinimide undergo condensation in the presence of basic catalysts (*56*). Similar condensations, remembering CLAISEN condensations of esters can occur also with imides which were added or which were formed by disproportionation during the base catalysed polymerization of caprolactam. The water which could be set free as a result of such a condensation consumes one equivalent of sodium caprolactam which is converted into the inactive sodium 6-aminocapronate. On the other hand the condensation products can take off directly another equivalent of the base because of their acidic nature.

The disappearance of the catalytic activity during the polymerization can be demonstrated by checking how fast a new equilibrium ratio monomer/polymer is being established when the temperature of the system is changed. After about 8 hours of heating to 220° the rate of equilibration was brought practically to stop (*67*), because no change of monomer content was observed when the temperature had been raised to 240° for 8 hours. The consumption of the catalyst was further followed by measuring the decrease of viscosity of solutions of polycaprolactam in a low molecular weight amide, e. g. diacetyl hexamethylene diamine in the presence of sodium caprolactam (*95*).

In full agreement with this explanation it was recognized that by a certain side reaction very strong cross links are formed during the polymerization. If a polymide was used as cocatalyst (a polymer styrene-methacryloyl caprolactam) the grafted polyamide side chains were joined in the very beginning of the polymerization and a crosslinked polymer resulted (*92*).

4. Inhibition effects

As compared with the hydrolytic polymerization, the base catalysed polymerization is essentially more sensitive towards impurities. The inhibition or retardation by traces of various foreign substances follows from the extremely low concentrations of the imide formed by the disproportionation reaction ranging about in the order of magnitude 10^{-2} mol-%, and from the high reactivity of the imide. The imides are the key intermediate in the polymerization mechanism and hence every substance which is able to react with imides acts either as an inhibitor or as a retarder of the base catalysed polymerization of caprolactam (*67*).

A very common impurity to be eliminated as far as possible is water which is converted in the presence of alkali caprolactam into alkali hydroxide; the latter in turn attacks immediately any imide, converting it in alkali carboxylate and amide (*76*). Thus, water destroys both active components, the base and the imide.

The polymerization is interfered by alcohols as well, due to their interaction with imides

$$-\text{CO}-\underset{|}{\text{N}}-\text{CO}- + \text{ROH} \rightarrow -\text{CO}-\text{NH}- + -\text{CO}-\text{OR}$$

Although there is not a direct simultaneous consumption of the base, even this possibility was admitted, assuming that alkylation of the amide group by the ester may occur (*25*). This seems however not to be correct, as the alkylation of amides produced by esters is limited to esters of aromatic acids only (*73*). Nevertheless, the esters might be accounted for a loss of basicity from the point of view of condensation reactions producing water and acidic products.

In accordance with the reaction mechanism proposed, primary and secondary amines behave as very effective inhibitors, due to their great reactivity towards imides.

With respect to the base, it is obvious, that any impurity which can reduce the basicity of the alkali caprolactam or completely neutralize the amide anions $[-\text{CO}-\text{N}-]^-$, i. e. any acid stronger than caprolactam, destroys completely the catalytic activity.

5. Molecular weight and distribution

According to varying conditions (temperature, reaction time, imide concentration, stabilizer etc.) the base catalysed polymerization may lead to products of a very wide range of molecular weights and types of distribution. It is possible to prepare without difficulty polymers ranging from a degree of polymerization 3 up to about 10^3.

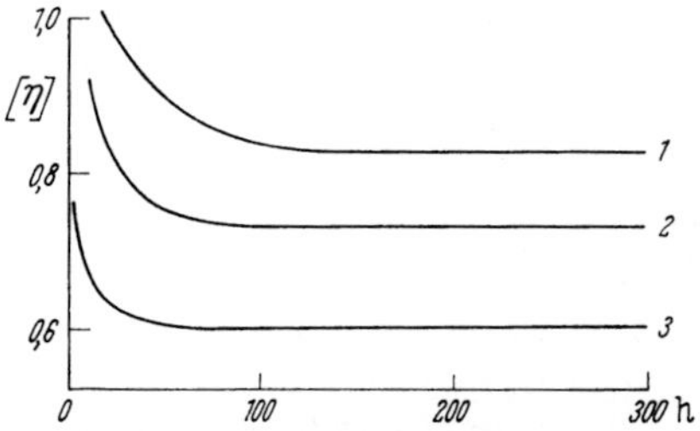

Fig. 2. Dependence of the intrinsic viscosity of polycaprolactam on the time of heating at 243°; c_{NaCL}=0,3 mol-% (*1*), 0,5 mol-% (*2*), 1,0 mol-% (*3*)

If the polymerization is performed in the presence of added imides, each polymer molecule contains in the average one imide group by means of which the endstanding cycle is attached (*68*). If the imide and sodium caprolactam are added in concentrations differing not more than one order of magnitude, the number average molecular weight equals the reciprocal imide concentration (*50*). When the concentration of the sodium salt surpasses essentially the imide concentration, imides formed by disproportionation play a relatively significant role and it is obvious, that the degree of polymerization is lower, than the reciprocal concentration of the initially added imide. The same effect on the molecular weight as is produced by imides can be achieved by the addition of low molecular open chain amides R-CO–NH–R′, which by taking part in the transamidation reactions introduce additional end groups.

After having reached the equilibrium ratio polymer/monomer the intrinsic viscosity shows its maximum value, depending on the temperature, concentration and sort of catalyst (*29, 49, 94*). By prolonged heating the intrinsic viscosity decreases and after a very sharp decrease during the first stage the viscosity approaches a certain limiting value which depends on the temperature and on the sort and concentration of catalyst (fig. 2). This decrease is a result of degradation reactions and changes in the distribution function (*29, 42, 43, 49*) (fig. 3).

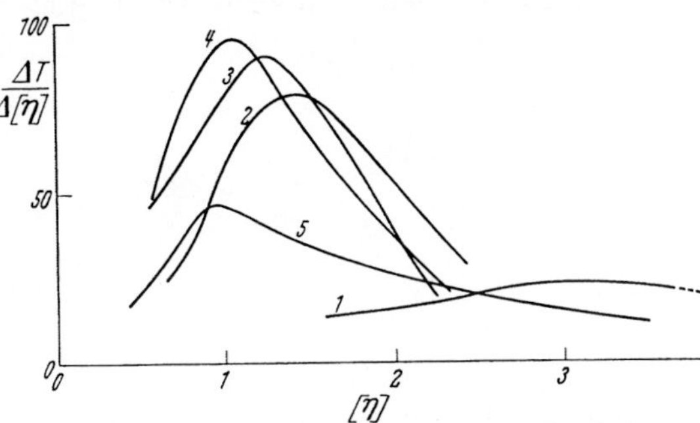

Fig. 3. Distribution curves of polycaprolactam during heating at 243° (c_{NaCL} = 0,3 mol-%); 1 h (*1*), 22 h (*2*), 50 h (*3*), 120 h (*4*), 500 h (*5*)

The initial distribution as determined by turbidimetric titrations of a polymer just formed is rather broad and by further heating it is converted slowly to a more homogenous statistical Flory-Schulz distribution (*27, 29, 49*); the increase of high molecular weight fractions by prolonged heating (curve 5 on fig. 3) may be explained by the formation of branched structures. In this redistribution of the polymer reactions can be involved which fit in the general transamidation scheme mentioned above. The fast anion–imide interaction, if applied to amide anions occurring anywhere inside the polymer chain gives rise to a redistribution reaction

CO ⁻N
N CO →
CO

→ CO–N
CO
N⁻
CO

Besides, the disproportionation reaction between amide groups and amide anions represents a redistribution, too, if linear amidic groups of the polymer are concerned (scheme (g)); in this case the number of particles remains constant, whereas when the disproportionation proceeds between a monomer molecule (or its anion) and an anionic group (or resp. amide group) inside the polymer chain the number of macromolecules is increased and the viscosity is decreased

$$\sim CO{-}NH \sim \; + \; {}^{-}N{-}CO \;(\text{ring}) \;\rightarrow\; \sim CO{-}N{-}CO \;(\text{ring}) \; + \; \sim NH_2$$

All these reactions are brought to stop after the basic amide anions have disappeared as a result of the side reactions. There is a good agreement in the period of the rapid degradation and redistribution as indicated by the sharp decrease of viscosity (fig. 2 and 3) and the time when the catalytic activity (basicity) expires (*49, 67*). Even after this time there remains a slow continuation of interchange reactions in which the remaining traces of basic and imide structures take part. In addition, perhaps also the less efficient groups like $-NH_2$, COO^- etc. formed in the course of the polymerization and further heating, could have a slight interchange activity (fig. 3).

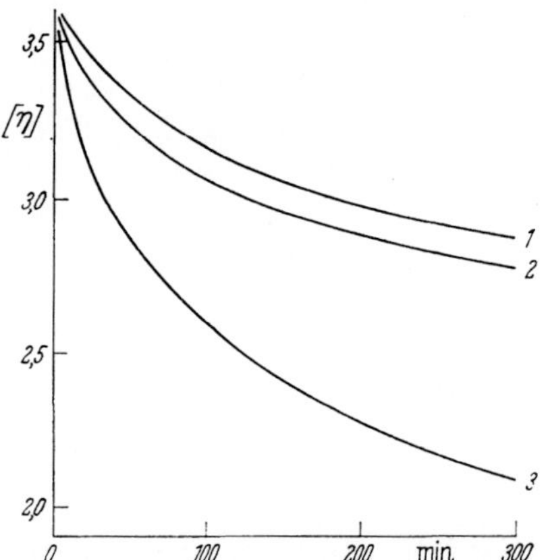

Fig. 4. Change of viscosity of polycaprolactam heated at 250°; initial polymer prepared with 0,06 mol-% NaCL + 0,10 mol-% AcCL (*3*), extracted with 1% aq. acetic acid, followed by extraction with water (*2*), extraction with water (*1*)

As any of the degradation reactions depends on the presence of a strong base, it is evident that neutralizing the base supports to the stabilization of the polymer. Thus, after having extracted the polymer either by water or by diluted acids, and after having dried the polymer, the decrease of the viscosity is much slower than in the case of non extracted polymer (*103*) (fig. 4). This makes it possible to apply the base catalysed polymerization for the production of polyamides for extruding and molding purposes.

The final molecular weight of the polyamide may be reduced, if desired, not only by increased initial imide concentrations (*30, 50*) or by addition of end group forming amides, but even by carrying out the polymerization at lower temperature. The region of limited solubility of the polymer in the reaction mixture is reached after a certain conversion, if the polymerization is performed below about 150°. The lower the temperature is, the more limited is the solubility of the polymer and the lower is the molecular weight of the fractions precipitated (*68*). At a temperature of 95° and 75° a polymer of an average degree of polymerization of about 20 and 10 respectively is formed. The separated polymer abstracts the imide from the liquid phase where the polymerization is taking place. Therefore the polymerization rate decreases and comes to a stop if all imide structures are incorporated into the precipitated polymer, irrespective of the overall conversion (*103*). By diluting the monomer by inert solvents the lowest possible oligomers can be obtained because of the further reduced solubility of the polymer and the possibility of carrying out the polymerization below the melting point of the caprolactam.

6. The polymer-monomer equilibrium

The equilibrium ratio polymer/monomer is the same in the base catalysed polymerization as in the hydrolytic process (*43*, *94*) and so is the equilibrium content of cyclic oligomers in both processes (*43*). Using the extremely fast basic polymerization it was made possible to follow the equilibrium in the range of very low temperatures, where it was hardly possible before. It was found that in polymers which were prepared below their melting point, the monomer content is essentially lower than would be expected from the extrapolation of the temperature dependence of the equilibrium monomer content obtained at higher temperatures (*94*) (fig. 5). This indicates that the crystaline fraction of the polymer does not participate in the equilibrium. It is possible to establish the concentration of the amorphous part independently, e. g. from density measurements. Assuming that the crystaline phase is monomer-free and that the extrapolated equilibrium concentration concerns the amorphous phase only (dotted curve), we can calculate the average monomer content in the samples prepared at low temperatures; experimental results are in a good agreement with these calculations.

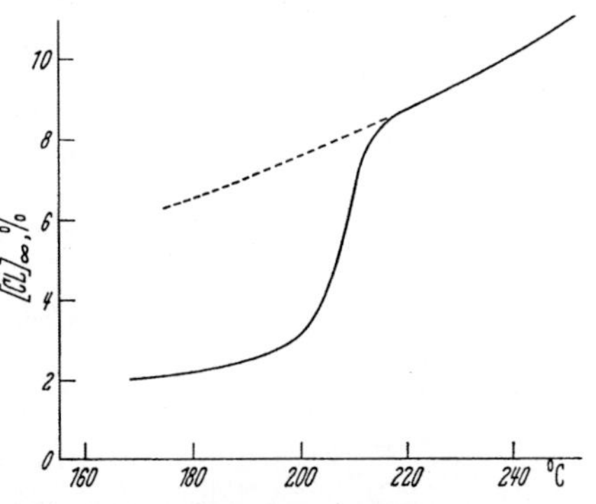

Fig. 5. Dependence of the equilibrium content of caprolactam $[CL]_{\infty}$ on the temperature

The fast establishement of the monomer-polymer equilibrium in the presence of basic catalysts makes it possible to arrange depolymerization experiments, if the monomer formed is distilled off from the heated polymer. For this purpose sodium and potassium carbonates have been found as the most convenient catalysts (*85*, *88*). They maintain the level of base concentration at a relatively low value and make it possible to depolymerize the polymer without important amounts of by-products. The rate of depolymerization is only very little affected by the molecular weight of the polymer (*88*). This indicates that the amino and carboxylic end groups occurring in current polycaprolactam samples do not contribute to the depolymerization. It is to emphasize that even mixed polyamides containing caprolactam, diamine and dicarboxylic acid units undergo a carbonate catalyzed depolymerization, leaving at last a residue of a nylon 66-type polyamide. Even in this case the rate of depolymerization is about the same as in pure caprolactam polymers (*85*). Hence the caprolactam units seem to be eliminated from the inside of the chain as easily as from the end groups with respect to the very fast base catalysed transamidation reactions

CO
N^{-} CO
N–CO
→
CO
N–CO + $^{-}$N–CO

7. Technological aspects

Using the system imide + base as catalyst, the polymerization of caprolactam increases over several orders of magnitude as compared with the previously known polymerization process. The polymerization below the melting point of the polymer can be, therefore, carried out without difficulty (*50*) by the catalysts mentioned above. The extraordinary low monomer content in the resulting polymer is valuable from the technological point of view. Moreover, the relatively low polymerization and crystalization enthalpy of caprolactam resp. polycaprolactam makes it possible to find out exact conditions to ensure the maintenance of temperature during the whole polymerization and crystalization process below the melting point of polycaprolactam even if the arrangement of the experiment is strictly adiabatic. In this way a solid block of polycaprolactam can be formed in several minutes in any large form of any shape (*98*). If the adiabatic conditions are maintained carefully, the block is uniform after the polymerization and crystalization has been completed, because all changes including the crystalization proceed at the same time anywhere in the polymeric material. The overall contraction caused both by crystalization (*26*) and polymerization is partly compensated by the thermal expansion during the adiabatic process. The internal tensions are therefore by far not so significant as in blocks prepared by the remelting of the polymer.

The high rate of polymerization which is especially enormous at temperatures above the melting point of the polymer where the polymer is formed within a few seconds, can be exploited in a continuous polymerization extruding process which make unnecessary to produce the polymer as an intermediate (*99*).

References

1. Arnold, Hoffman & Co., Inc.: Brit. pat. 731,294 (2. 10. 1952).
2. Arnold, Hoffman & Co., Inc.: Brit. pat. 754,944 (24. 2. 1954).
3. Badische Anilin & Soda Fabr., A. G.: Brit. pat. 820,607 (23. 9. 1959).
4. Barnes, C. E., W. O. Jr. Ney and W. R. Nummy: U.S. pat. 2,711,398 (21. 6. 1955).
5. —, —, — and R. J. Warren: U.S. pat. 2,809,958 (15. 10. 1957).
6. —, W. R. Nummy and W. C. Ney: U.S. pat. 2,806,841 (17. 9. 1957).
7. Benson, R., and T. Cairns: J. Am. Chem. Soc. **70**, 2115 (1948).
8. Berthold, G. H.: U.S. pat. 2,727,017 (13. 12. 1955).
9. —, and R. N. Lewis: U.S. pat. 2,857,364 (21. 10. 1958).
10. BIOS Final Report No. 1472.
11. Blacher, C.: Ber. dtsch. chem. Ges. **28**, 432 (1895).
12. Black, W. B., and H. G. Clark: U.S. pat. 2,912,415 (10. 11. 1959).
13. Bobingen, A. G.: Germ. pat. 904,948 (23. 10. 1942).
14. Champetier, G., and H. Sekiguchi: C. R., 249,108 (1959).
15. — — Lecture on the IUPAC Symposium on Macromolecular Chemistry, 1960, Moscow, Section II. p. 195.
16. Chrzczonowicz, S., and S. Dyniewski: Práce Głównego Inst. Wlok. **5**, 1 (1951).
17. —, and J. Moniewski: Zeszyty Nauk Politech. Łódz, Chem. **1**, 77 (1954).
18. — Zeszyty Nauk Politech. Łódz, Chem. **3**, 93 (1955).
19. — Zeszyty Nauk Politech. Łódz, Chem. **5**, 65 (1957).
20. — Pol. pat. 41,536 (20. 7. 1958).
21. —, and M. Wlodarczyk: IUPAC Symposium on Macromol. Chem. Wiesbaden 1959, Short Communications III-C-5.
22. —, B. Ostaszewski and M. Wlodarczyk: IUPAC Symposium on Macromolecular Chem., 1960, Moscow, Materials Section II, p. 497.
23. Crowther, M.: U.S. pat. 2,734,043 (7. 2. 1956).
24. Curtius, T.: Ber. dtsch. chem. Ges. **23**, 3023 (1890).
25. Čefelín, P., E. Šittler and O. Wichterle: Collection Czechoslov. Chem. Communs. **24**, 3287 (1959).
26. —, J. Trekoval and Z. Drbálek: Collection Czechoslov. Chem. Communs. **24**, 2890 (1959).
27. Gordienko, A., W. Griehl u. H. Sieber: Faserforsch. u. Textiltech. **6**, 105 (1955).

28. Griehl, W.: Faserforsch. u. Textiltech. **6**, 260 (1955).
29. — Faserforsch. u. Textiltech. **7**, 207 (1956).
30. —, u. S. Schaaf: Makromol. Chem. **32**, 170 (1959).
31. Hall, H. K., jr.: J. Am. Chem. Soc. **80**, 6404, and 6412 (1958).
32. — J. Am. Chem. Soc. **82**, 1209 (1960).
33. Hamann, A.: D.R. pat. 748,253 (11. 6. 1938).
34. — Germ. (East) pat. 8188 (16. 1. 1951).
35. — Germ. (East) pat. 7667 (24. 5. 1952).
36. — Germ. (East) pat. 7657 (15. 7. 1952).
37. — Germ. (East) pat. 11,086 (5. 1. 1956).
38. — Brit. pat. 762,443 (9. 11. 1954).
39. — Brit. pat. 762,444 (9. 11. 1954).
40. — Brit. pat. 762,927 (25. 10. 1955).
41. — Faserforsch. u. Textiltech. **9**, 351 (1958).
42. Hanford, W. E., and R. M. Joyce: J. Polymer Sci. **3**, 167 (1948).
43. Heikens, D.: Makromol. Chem. **18/19**, 62 (1956).
44. Hubert, E.: Germ. pat. 906,512 (26. 4. 1942).
45. —, u. A. Hamann: Germ. (East) pat. 5067 (25. 10. 1942).
46. Joyce, R. M.: Swed. pat. 99,037 (7. 6. 1939).
47. —, and D. M. Ritter: U.S. pat. 2,251,519 (5. 8. 1941).
48. Koch, Th.: U.S. pat. 2,622,076 (16. 12. 1952).
49. Králíček, J., and J. Šebenda: J. Polymer Sci. **30**, 493 (1958).
50. —, — Z. Zadák and O. Wichterle: Chem. prům. (in print).
51. Lautenschlager, H., H. Friedrich, W. Schmidt, and K. Dachs: U.S. pat. 2,907,755 (5. 9. 1957).
52. Mighton, H. R.: U.S. pat. 2,647,105 (28. 7. 1953).
53. Müller, A., u. O. Bleier: Monatsh. f. Chem. **50**, 399 (1929).
54. Ney, W. O. jr., W. R. Nummy and C. E. Barnes: U.S. pat. 2,638,463 (7. 12. 1951).
55. —, and M. Crowther: U.S. pat. 2,739,959 (27. 3. 1956).
56. Plešek, J.: Résumés des Communications XVIe Congrés IUPAC Paris, Tome II, 164 (1957).
57. Rakshit, J. T.: J. chem. Soc. (London) **103**, 1557 (1913).
58. Rothe, M.: J. Polymer Sci. **30**, 227 (1958).
59. Ruzicka, L.: Helv. Chim. Acta **4**, 472 (1921).
60. Saunders, J.: J. Polymer Sci. **30**, 479 (1958).
61. Schaaf, S.: Faserforsch. u. Textiltech. **10**, 224 (1959).
62. — Faserforsch. u. Textiltech. **10**, 257 (1959).
63. — Discussion on the International Symp. on Macromolecular Chem. Prague 1957. Collection Czechoslov. Chem. Communs. **22**, 292 (1957)
64. Schlenker, F., u. H. Stark: Germ. (East) pat. 2719 (26. 4. 1942).
65. Šebenda, J.: Kunststoff-Rdsch. **4**, 245 (1957).
66. —, and J. Králíček: Czech. pat., 93016 (17. 10. 1957).
67. —, — Collection Czechoslov. Chem. Communs. **23**, 766 (1958).
68. —, — IUPAC Symposium on Macromol. Chem. 1959, Wiesbaden, Short Communications III - C - 6 (Weinheim: Verlag Chemie G.m.b.H.)
69. Šostakovskij, N. F., N. A. Nedzychovskaja and M. N. Zelenskaja: Izvest. Akad. Nauk SSSR., Otdel. Khim. Nauk **1952**, 682.
70. Špitalnyj, A. S., and N. S. Jabločnik: Zhur. Obshcheî Khim. **28**, 3282 (1958).
71. —, M. A. Špitalnyj and N. S. Jabločnik: Zhur. Priklad. Khim. **32**, 617 (1959).
72. —, — Zhur. Obshchei Khîm. **29**, 1285 (1959).

73. THOMAS, P. R.: IUPAC Symposium on Macromol. Chem. 1959 Wiesbaden, Short Communications IV - B - 6.
74. TITHERLEY, A. W.: J. Chem. Soc. (London) **81**, 1520 (1902).
75. — J. Chem. Soc. (London) **85**, 1673 (1904).
76. —, and L. STUBBS: J. Chem. Soc. (London) **105**, 299 (1914).
77. TOMASZEWSKI, J., and S. CHRZCZONOWICZ: Zeszyty Nauk Politech. Łódz, Chem. **6**, 57 (1957).
78. TURSKA, E., and A. BRODA: Zeszyty Nauk Politech. Łódz, Włókiennictvo **4**, 49 (1954).
79. VOSS, W.: Chem. Tech. (Berlin) **1**, **111** (1949).
80. WALLACH, O.: Liebigs Ann. Chem. **324**, 292 (1902).
81. — Liebigs Ann. Chem. **343**, 40 (1905).
82. WICHTERLE, O., and J. ŠEBENDA: Czech. pat. 87163 (21. 10. 1952).
83. —, — Czech. pat. 87814 (24. 4. 1955).
84. — Faserforsch. u. Textiltech. **6**, 237 (1955).
85. —, J. ŠEBENDA and J. KRÁLÍČEK: Faserforsch. u. Textiltech. **6**, 563 (1955).
86. —, J. KRÁLÍČEK and J. ŠEBENDA: Czech. pat. 89,233 (16. 2. 1956).
87. —, and J. ŠEBENDA: Collection Czechoslov. Chem. Communs. **21**, 312 (1956).
88. —, J. KRÁLÍČEK, J. ŠEBENDA and A. CHALUPNÁ: Faserforsch. u. Textiltech. **7**, 393 (1956).
89. — Lecture on the Meeting of the German Chem. Soc., Leipzig, October 1956.
90. — Collection Czechoslov. Chem. Communs. **22**, 288 (1957).
91. —, and J. ŠEBENDA: Collection Czechoslov. Chem. Communs. **22**, 1353 (1957).
92. —, and V. GREGOR: J. Polymer Sci. **34**, 309 (1958).
93. —, J. KRÁLÍČEK and J. ŠEBENDA: Collection Czechoslov. Chem. Communs. **24**, 755 (1959).
94. —, E. ŠITTLER and P. ČEFELÍN: Collection Czechoslov. Chem. Communs. **24**, 2356 (1959).
95. —, —, — Communication on the IUPAC Symposium on Macromol. Chem., 1960, Moscow. Doklady i avtoreferaty, Sekcija III, 380.
96. — Makromol. Chem. **35**, 174 (1960).
97. —, E. ŠITTLER and P. ČEFELÍN: Collection Czechoslov. Chem. Communs. (in print).
98. —, J. KRÁLÍČEK and J. ŠEBENDA: Czech. pat. 97,333 (23. 11. 1957).
99. —, J. ŠEBENDA and J. KRÁLÍČEK: Czech. pat. 97,332 (23. 11. 1957).
100. YODA, N., and A. MIYAKE: J. Polymer Sci. **43**, 117 (1960).
101. YUMOTO, H., and N. OGATA: Bull. Chem. Soc. Japan **31**, 907 (1958).
102. —, — Bull. Chem. Soc. Japan **31**, 913 (1958).
103. Unpublished experiments from our laboratory.

Sachverzeichnis

Die Stichworte sind in der Sprache der Beiträge wiedergegeben.

FORTSCHRITTE DER HOCHPOLYMEREN-FORSCHUNG

ADVANCES IN POLYMER SCIENCE

HERAUSGEGEBEN VON

J. D. FERRY
MADISON

C. G. OVERBERGER
NEW YORK

G. V. SCHULZ
MAINZ

A. J. STAVERMAN
LEIDEN

H. A. STUART
MAINZ

2. BAND

MIT 95 ABBILDUNGEN

Springer-Verlag Berlin Heidelberg GmbH

ISBN 978-3-540-02626-6 ISBN 978-3-540-37047-5 (eBook)
DOI 10.1007/978-3-540-37047-5

Ursprünglich erschienen bei Springer Verlag oHG Berlin Göttigen Heidelberg 1961.

Brühlsche Universitätsdruckerei Gießen

Inhalt des 2. Bandes

1. Heft

Seite

2. Heft

3. Heft

4. Heft